THE U.S. 37-MM GUN IN WORLD WAR II

CASEMATE | ILLUSTRATED | SPECIAL

CASEMATE | ILLUSTRATED | SPECIAL

THE U.S. 37-MM GUN IN WORLD WAR II

CHARLES C. ROBERTS, JR.

CISS0015

Published in the United States of America and Great Britain in 2023 by
CASEMATE PUBLISHERS
1950 Lawrence Road, Havertown, PA 19083, USA
and
The Old Music Hall, 106–108 Cowley Road, Oxford OX4 1JE, UK

© Charles C. Roberts Jr., 2023

Print Edition: ISBN 978-1-63624-252-1
Digital Edition: ISBN 978-1-63624-253-8

A CIP record for this book is available from the British Library.

Design by Mel Gallipeau, www.melgallipeau.com
Printed and bound in the Czech Republic by FINIDR s.r.o.

For a complete list of Casemate titles, please contact:
CASEMATE PUBLISHERS (US)
Telephone (610) 853-9131
Fax (610) 853-9146
Email: casemate@casematepublishers.com
www.casematepublishers.com

CASEMATE PUBLISHERS (UK)
Telephone (0)1226 734350
Email: casemate-uk@casematepublishers.co.uk
www.casematepublishers.co.uk

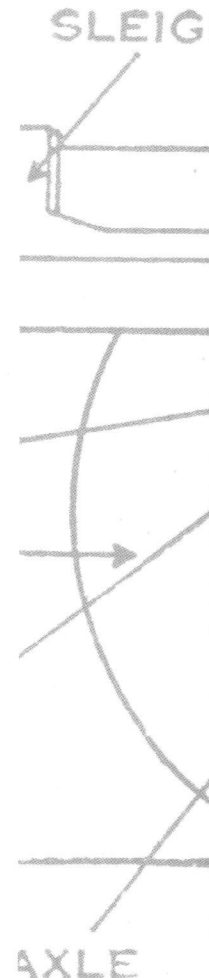

Page 2 image: Prior to the U.S. entry into World War II, Erie Proving Grounds in
  Ottawa County, Ohio was testing the M3 37-mm antitank gun. Guns from various
  manufacturers and of various designs were lined up at various positions and tested for
  compliance to specifications and reliability. (U.S. Army)
Title page image: Restored M5A1 Stuart light tank in a reenactment. (Roberts Collection)
Contents page images: 37-mm M3 gun on M4 carriage elevation drawing.
  (FM 23-70) & M8 armored car left front view. (Roberts Collection)

Special thanks to the following for help in reviewing this book: Michael Krizsanitz, Lydia
Roberts and Jeff Wszolek.

Contents

37-mm antiaircraft gun, Solomon Islands. (Wikimedia Commons)

| Timeline of Events

1862: The Lieber Code, defining the conduct of war, is signed into law by American President Abraham Lincoln.

1868: St. Petersburg Declaration limits explosive charges in projectiles. The United States is not a signatory.

1899: The Hague Convention, based largely on the Lieber Code, reaffirms the ban. The United States is a signatory.

April 6, 1917: The United States officially enters World War I. Model M1916 37-mm gun sees continuous service during the war.

1919–1940: M1916 37-mm gun remains in infantry service but as war looms is gradually replaced by M3 37-mm gun.

December 1938: Development testing by the U.S. Ordnance Department results in the T10 gun and T5 gun carriage, culminating in the M3 antitank gun.

May 1940: Production of the M2A4 begins, the first tank entering service with the 37-mm gun M5.

July 1940: Plans are developed for the first medium tank, the Medium Tank M3 "Lee", known as the "Grant" by the British (because of a different turret).

March 11, 1941: U.S. Congress passes the Lend-Lease Act into law. Military supplies to the UK, Soviet Union, China, France, Brazil, and other Allied countries commence.

July 1941: The Ordnance Department publishes specifications for the T17E1 Stag-hound, some 3,800 of which are purchased by the British.

M6 37-mm gun motor carriage left front view. (TM 9-750A)

December 7, 1941: Japan attacks American naval base at Pearl Harbor.

December 11, 1941: The United States officially enters World War II.

December 1941: The 37-mm M3 towed antitank gun is the first M3 to see combat during the defense of the Philippines.

April 1942: In the Army organization chart, each infantry battalion has an antitank platoon with four 37-mm towed guns hauled using Jeeps or ¾-ton trucks. Divisional artillery battalions have six 37-mm towed guns and combat engineering battalions have nine guns towed by M2 halftracks. Divisional headquarters company has four 37-mm towed guns.

Mid-1942: The British Eighth Army in North Africa relegates its M3 Stuarts to non-combat roles. The British had initially ordered 700 Stuart tanks, known by them as "Honeys."

January 29, 1943: Carriage M4 designation changes to Carriage M4A1. Gun carriage production is ramped up.

August 1942–February 1943: M3 Stuart light tank provides solid service during the Guadalcanal campaign.

March 1943: Production commences on the M8 Greyhound armored car, followed in quick succession by the T18 Boarhound, M38 Wolfhound, and the T17E1 Staghound.

Mid-1943: Unable to keep up with German armor developments, the M3 becomes ineffective in Europe and is redeployed to the South Pacific where it is effective against Japanese light armor. The PT boat arm of the U.S. Navy starts installing AN-M4 37-mm guns on the foredecks of torpedo boats.

July 1945: War production effectively ceases.

| Introduction

The 1899 Hague Convention was the first significant multinational treaty that set forth the conduct requirements for war. It was based on President Lincoln's Lieber Code that codified law relating to the behavior of Union Soldiers during the American Civil War. The Hague dictates were welcomed and adopted by military establishments in many countries. The 1899 Hague Convention set forth limitations on the size of weapons that would fire explosive shells.

Early on, bullets that would explode upon striking a soldier were being developed. This was considered unnecessary and inhumane since it was thought that a bullet without explosives would do the job. An exploding projectile with a weight of a minimum of 0.88 lb was allowed by the convention. Consequently, gun developers settled on the 37-mm gun as a weapon that would meet the criterion of the Hague but be light and lethal enough to be used in battle. One of the first light guns was the Model 1916 37-mm designed and built by the French using a projectile weighing approximately 1 lb. It was used extensively in World War I by the French and the American forces.

Between World War I and World War II, several countries developed or adopted the 37-mm gun. Bofors of Sweden developed a very capable 37-mm antitank gun in the 1930s that was sold to or licensed with many countries. Bofors of Sweden was one of the early developers of a 37-mm antitank gun. It was well designed, easy to operate, efficient and supplied with reliable ammunition. It was utilized by many countries in World War II including Finland, Poland, and Britain. The Germans developed the Pak 36 37-mm antitank gun that was adopted by their military and was also well received by many countries. The United States was behind in the development of an antitank gun but relied on the German design as a basis for development.

By the mid-1930s, the U.S. Ordnance Department had designed the M3 37-mm gun and M4 carriage, resulting in a towed antitank gun—the first antitank gun in the U.S. Army. This gun proved effective at the beginning of World War II, but as German armor protection increased, it was unable to penetrate the frontal armor of many German tanks and was relegated to lesser roles. However, the gun proved effective against the Japanese tanks and Japanese strongpoints in the Far East.

The American military used the 37-mm gun on several production and experimental armored vehicles, including the M2A4 Light Tank, the M3 Lee Medium Tank, the M3 Stuart Light Tank, the M5 Stuart Light Tank, the M22 Locust Airborne Tank, the M8 Armored Car, the T17E1 Staghound Armored Car and the M3A1E3 Scout Car. The gun was also used on several production and experimental non-armored vehicles, including the M6 37-mm Gun Motor Carriage, the T2 37-mm Gun Motor Carriage, the T8 37-mm Gun Motor Carriage, the T14 37-mm Gun Motor Carriage, the T33 37-mm Gun Motor Carriage, the M3 towed gun on the M4 carriage, the P-39 Airacobra, P-63 Kingcobra, and on selected naval vessels. Despite its small size, the American M3 37-mm gun served throughout the war, on many vehicles and performed exactly as designed.

M1916 37-mm gun and crew.
(Wikimedia Commons)

| 1 Development of the 37-mm Gun

In 1863, the Russians had developed a musket ball that would explode when hitting a hard target such as ammunition dumps or ammunition wagons. In 1867, the Russians perfected an improved version of the exploding musket ball that would detonate when impacting soft targets (i.e., humans). The resulting exploding musket ball would be horrid to a soldier when a single standard musket ball would be sufficient. In order to prevent an out-of-control arms race with highly destructive weapons, the Russians decided to set up a conference that would limit the proliferation of such weapons.

Prince Alexander Gorchakov, on behalf of the Russian government, decided to convene a conference with the purpose of eliminating or reducing the production of these disastrous weapons. At the invitation of Prince Gorchakov, a conference of representatives from several European governments met in St. Petersburg, Russia, in December 1868, with the purpose of revising the "rules of war." Attendees were represented from the following nations: Austria-Hungary, Bavaria, Belgium, Denmark, France, British Empire, Greece, Italy, Netherlands, Portugal, Prussia, Russia, Sweden-Norway, Switzerland, Ottoman Empire and Württenberg. The United States was not considered a world power and did not take part in any aspect of the conference. The convention delegates agreed that the purpose of war was to diminish the capability of the opposing military force. They also agreed that these small-caliber explosive projectiles should be banned as they tended to increase injury rates to the combatants and prolong suffering.

The attendees also agreed in the St. Petersburg Declaration of 1868 that a projectile weighing less than 0.88 lb with an explosive charge would be banned. The reason was that smaller exploding projectiles would kill only one soldier, and that a bullet would do the same, negating the need for such inhumane weapons.

The Hague Convention of 1899 was the first that resulted in multilateral treaties between countries to address the conduct of warfare. The treaties were based on the Lieber Code. The Lieber Code, entitled *Laws and Usages of War*, was published by Professor Francis Lieber in August 1862 at the request of Major General Henry W. Halleck, General in Chief of the Army of the United States. The code was signed into law by President Abraham Lincoln and issued to the Union Forces in April 1863, during the American Civil War. This was the first codified law that set out regulations for behavior in times of martial law: protection of civilians, punishment for deserters, protection of prisoners of war, handling of hostages, punishment for pillaging, treatment of spies, organizing a truce, punishment for assassinations of soldiers and prisoner exchange. The code was widely circulated among foreign nations and was adopted by many military establishments. Much of the wording in treaties agreed upon at the 1899 Hague Convention was borrowed from the Lieber Code. The 1899 Hague Convention reaffirmed the ban on explosive projectiles weighing less than 0.88 lb. The United States participated in the 1899 conference.

Model M1916 37-mm Gun in World War I

The limitation in weight of a projectile that could explode gave birth to the 37-mm projectile. The model 1916 37-mm gun was the smallest practical gun allowed by the 1899 Hague Convention that could utilize explosive projectiles weighing approximately 1 lb and having a shape consistent with projectile design at that time. The 37-mm gun was attached to a tripod with wheels so that the gun could be towed by two men or lifted by four men. The breech was of interrupted thread design, a smaller version of the French 1897 75-mm gun. When in place, the gun could be crewed by two men: a loader and a gunner. The gun was equipped with sights for direct fire (APX telescopic sight) and indirect fire (quadrant sight). When being towed, an ammunition cart was attached to the gun carriage carrying fourteen 16-round boxes of ammunition. The Mark II explosive shell used by the U.S. weighed approximately 1.5 lb with a bursting charge of 0.06 lb of TNT (trinitrotoluene). During World War I, this gun was designated 37-mm M1916 in American service. Direct fire using this weapon was somewhat lacking since the crew had to be in the line of sight of the target and thus be vulnerable to enemy fire. It worked better as an indirect fire weapon, much like a mortar. This gun was fitted to the M1917 light tank that was mass produced by the U.S. but never saw combat during World War I.

Between World War I and World War II, infantry companies were equipped with Model M1916 37-mm guns but, by 1941, most were relegated to storage in favor of the M3 37-mm gun. Some were still used, as there was a shortage of the M3 guns at the beginning of the war. The gun was also used for training as a sub-caliber weapon on larger guns in order to conserve ammunition. In January 1942 two such guns were used against the Japanese in the defense of the Philippines.

Summer training at Fort Howard with the 70th Tank Battalion using the M1916 37-mm gun on a carriage, circa 1940. (C. Roberts, *Armored Strike Force: The Photo History of the 70th Tank Battalion in World War II*)

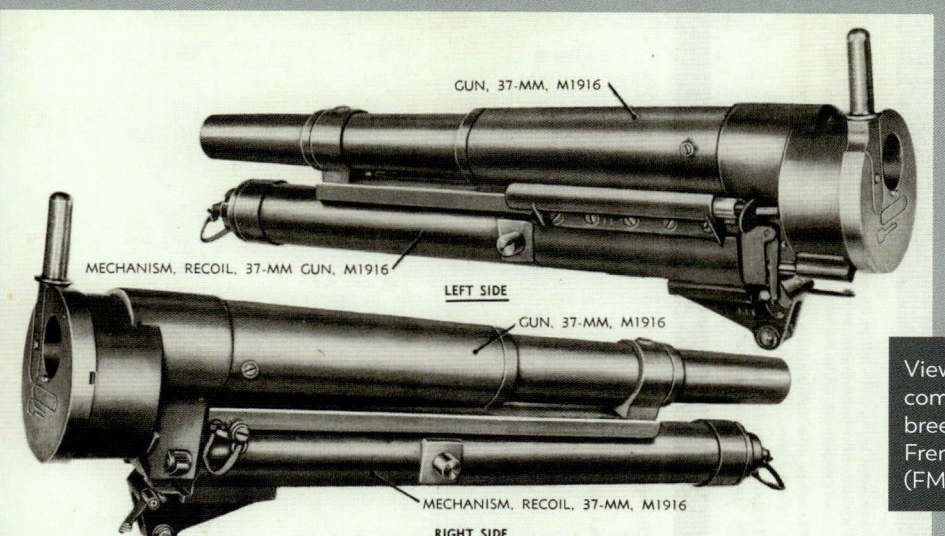

View of the M1916 37-mm gun components. The interrupted thread breech design was similar to the French 1897 75-mm guns of that era. (FM 23-75)

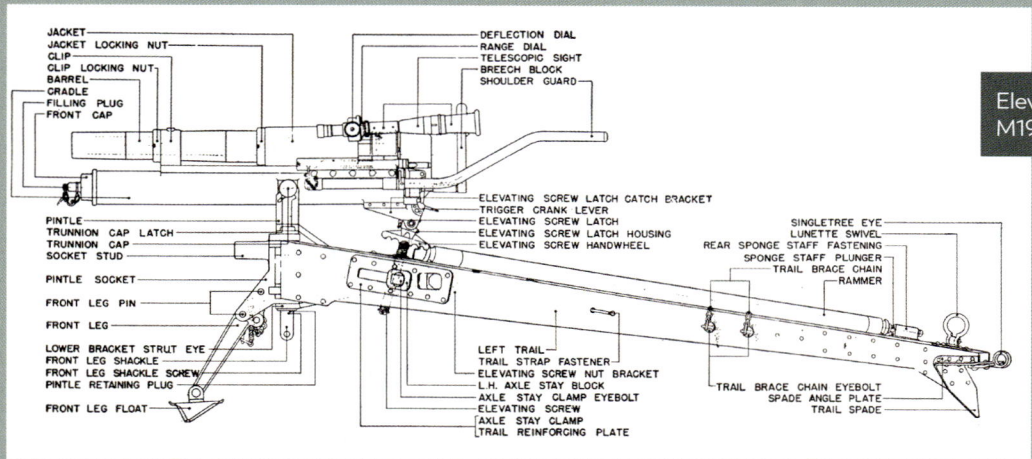

Elevation drawing of the M1916 37-mm gun. (FM 23-75)

Training with the 70th Tank Battalion circa 1940 showing the M1916 37-mm gun on a carriage. (C. Roberts, *Armored Strike Force: The Photo History of the 70th Tank Battalion in World War II*)

Finnish 37-mm Bofors antitank gun. (Wikimedia Commons)

Swedish Bofors 37-mm Antitank Gun

The Bofors Forging Company was established around the mid-1600s in Sweden with two iron bar forging hammers. By the late 1800s, Bofors became Sweden's largest manufacturer of rolled bar and Lancashire iron. Bofors produced a kind of steel that was especially suitable for the manufacture of cannons using the open-hearth type furnaces. The first foreign order came from Switzerland in 1888 for twenty-eight 12-cm guns. Alfred Nobel, the famous inventor, purchased the company in 1894 and brought money and fresh new ideas to the company. He emphasized manufacturing fewer products but of exceptional quality. By the end of World War I, the company enlarged the steel plant, enlarged the machine shops, manufactured gunpowder, developed an explosive factory and laid out a proving ground, all at the same location. As World War II approached, there was a huge increase in industrial production of armaments.

The Finnish Army manual gives the following statistics for armor penetration values of the 37-mm Bofors—all against 60° angle from horizontal—as follows:

328 yards = 1.57-inch penetration

547 yards = 1.29-inch penetration

1,093 yards = 0.71-inch penetration

Stridsvagn L-60 tank with Bofors 37-mm, (Roberts Collection)

The Royal Swedish Arms Commission was interested in developing an antitank gun that could deliver a 1.54-lb projectile at 2,600 ft/sec. Bofors stepped up in 1934 with a 37-mm monobloc barrel on a split trail carriage equipped with a 5-mm-thick gun shield and rubber wheels. The gun was also used as the main tank gun on tanks such as that on the Swedish Stridsvagn L-60.

Initial production of the Bofors was sent to the Netherlands. Later Poland, Finland, Britain and Denmark bought several guns and licenses. The Spanish Republicans bought 20–30 guns which were put to use during the Spanish Civil War—the first usage in combat. The gun was effective against the opposing Panzer I tanks, T-26s and CV33 tanks. Poland purchased 300 guns from Bofors from 1936 to 1939 and acquired a license to manufacture. When the Germans attacked in 1939, the Polish version of the gun (wz.36) knocked out many panzers, especially at the battle of Mokra.

Sweden adopted the Bofors 37-mm gun in 1937 as the 37-mm Infanterikanon M/43 and the 37-mm Pansarvärnskanon M/38 for use by infantry battalions that had antitank platoons equipped with two of the guns. The gun was installed in the Swedish Stridsvagn L-60 light tank shown on the opposite page. Denmark acquired a license to produce the Bofors gun in 1937 calling it the 37-mm Fodfolkskanon model 1937. During the German invasion of 1940, it performed well, knocking out several of the early-war lightly armored German tanks. In October 1938, Finland adopted the Bofors 37-mm gun as the new standard for their military, with guns being manufactured under license in two factories. When the Soviet Union invaded Finland in November 1939, the Bofors 37-mm gun was effective against Soviet tanks, including the T-26 and BT. Several enemy tanks were knocked out which solidified the weapon's reputation as a tank killer at that time. Romanian units used the Bofors 37-mm gun in Operation *Barbarossa* on the side of Germany, but it proved to be increasingly ineffective against the improved Soviet armor of the T-34 and KV-1.

The Bofors 37-mm was considered the best antitank gun in the 1930s. It had reliable ammunition and a high rate of fire. It could penetrate most of the tank armor at that time and was easily handled or adapted to military vehicles. It had a muzzle brake which was mounted at the end of the barrel and redirected gases toward the rear to reduce the recoil force.

Specifications: Bofors 37-mm Antitank Gun

Caliber	37 mm
Barrel length	68 in (incl. muzzle brake)
Weight in firing position	816 lb
Traverse	26°
Elevation	–10° to +25°
Max rate of fire	30 rpm
Practical rate of fire	12 rpm
Muzzle velocity	2,723 ft/sec
Max range	2.8 miles
Max practical range vs. tank	0.56 miles
Max practical range vs. infantry position	0.93 miles

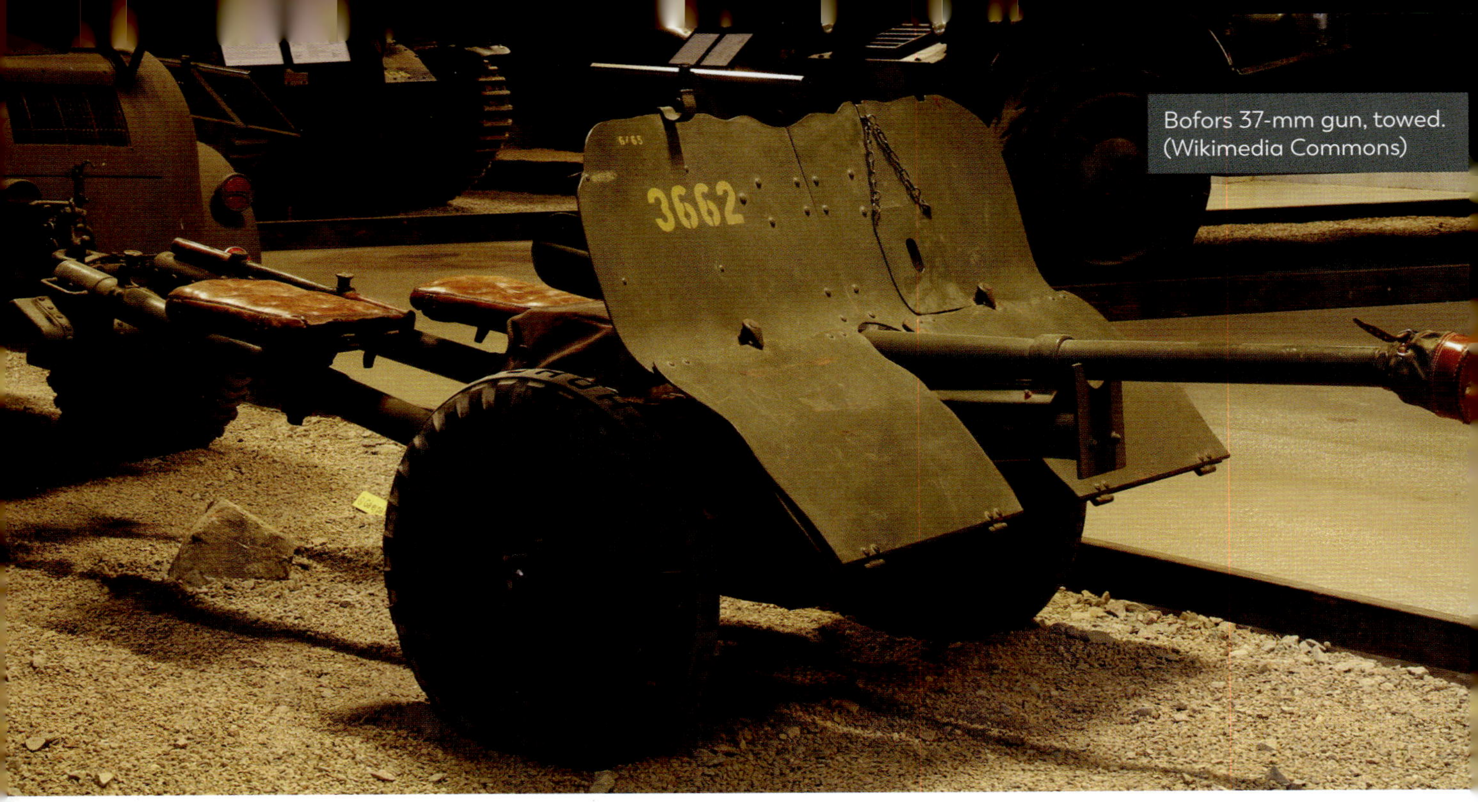

Bofors 37-mm gun, towed. (Wikimedia Commons)

German Pak 36 37-mm Antitank Gun

The Pak 36 (Panzerabwehrkanone 36) was the antitank gun used by the Wehrmacht Panzerjäger (antitank units) from 1936 to 1942. It was designed in 1933 by the Rheinmetall Corporation and issued to units in 1936. As a rival to the Bofors, the Pak 36 was exported to several countries and copied by the Soviet Union and Japan. It served in the Spanish Civil War and was effective against the light tanks of that conflict. At the beginning of World War II, it proved effective against Polish tanks but eventually became ineffective against French Char B-1 and British Matilda II tanks. During Operation *Barbarossa*, the Pak 36 destroyed many of the Soviet light tanks but not the T-34 and KV-1 tanks. In 1943, a shaped charge projectile was added (Stielgranate 41) which gave the gun the enhanced capability for defeating any enemy tank on the battlefield. However, the short range of this ammunition made the crew vulnerable to enemy fire. German paratroopers utilized the gun because of its light weight and maneuverability.

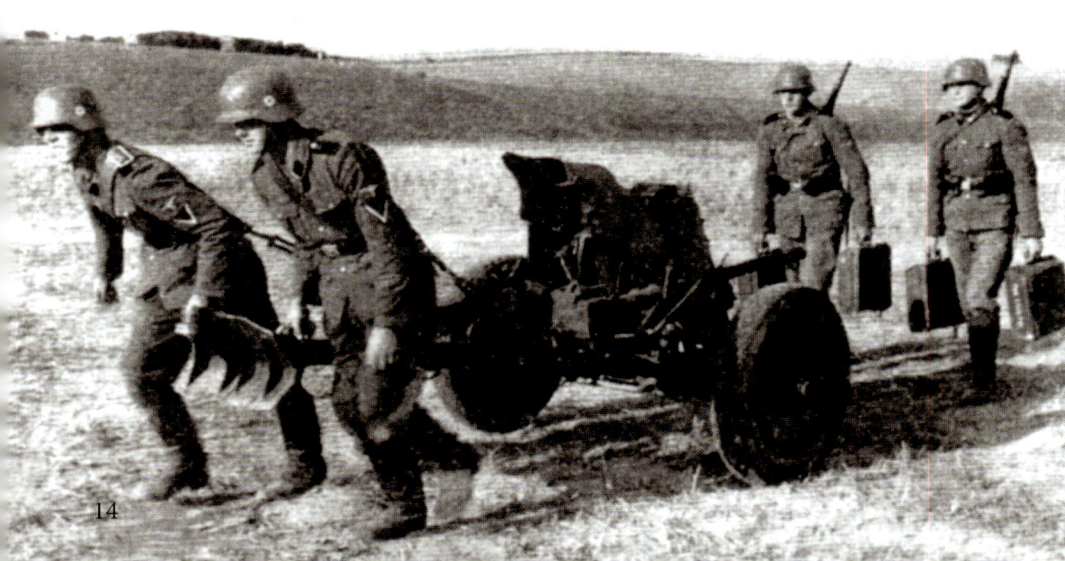

Pak 36 and crew. (TM-E 30-451)

Specifications: Pak 36 37-mm Antitank Gun

Caliber	37 mm
Weight	990 lb
Gun tube length	5 ft 5 in
Width	5 ft 5 in
Height	3 ft 10 in
Cartridge length	9.8 in
Cartridge diameter	1.54 in
Elevation	−5° to +25°
Rate of fire	13 rpm
Muzzle velocity	2,500 ft/sec
Maximum range	5,997 yd

Pak 36. (Roberts Collection)

Projectiles with a tungsten core were manufactured and resulted in increased penetration of steel targets. Tungsten weighs about twice as much as steel so the kinetic effect at impact is greater. This new projectile could penetrate the side armor of a T-34, but only at a very close range, making the crew vulnerable to counterattack. Despite the short comings of the Pak 36 in defeating the T-34 Russian tank, it remained the standard antitank weapon for many German units until 1942. The gun was light, had a high rate of fire, was easy to deploy and conceal. By mid-1941, the Pak 36 was being replaced with the 50-mm Pak 38 and was removed from carriages for application in other vehicles. The Pak 36 was installed in the Sonderkraftfahrzeug (Sd.Kfz.) 251 halftrack for use as a light anti-armor support vehicle. The Pak 36 was given to the Romanian Army but proved inadequate for stopping the Soviet advances. The Pak 36 also served in the armies of Italy, Finland, Hungary, China and Slovakia. Antiaircraft versions of the 37-mm gun were produced such as the Flak 18 and Flak 36. (Flak: abbreviated from *Flugabwehrkanone*—aircraft defense cannon)

German 3.7-cm Flak 18. This is the antiaircraft version of the 37-mm gun. (TM-E 30-451)

Pak 36 with Stielgranate 41 developed in 1941 with production ending in early 1943. (TM-E 30-451)

Sd.Kfz. 251 chassis with Flak 36 (37-mm AA gun) mounted (TM-E 30-451)

Japanese Type 94 37-mm Antitank Gun

The Type 94 37-mm antitank gun was used in the Second Sino-Japanese War and World War II.

The gun was developed by the Japanese Imperial Army and designated model 2594 (first year of manufacture, imperial year calendar) or 1934 (western calendar). Approximately 3,400 guns were produced and future models were designated Type 1 37-mm antitank gun. The organization of the combat infantry regiments required four guns with 11 soldiers crewing each gun.

The breech had a semi-automatic shell ejection system that would eject the spent shell after the gun recoiled. The gun could be disassembled into four packs weighing less than 220 lb and could fire high-explosive or armor-piercing rounds. The gun was equipped with a straight telescopic sight. The standard armor-piercing projectile could penetrate 1.7 inches of armor at 500 yards. Additional efforts were made by the Japanese Army Technical Bureau to increase muzzle velocity, thereby increasing penetration performance. The Type 94 37-mm AT gun was effective against the Soviet BT tanks but not against the American tanks, such as the M4 Sherman. The Type 1 47-mm antitank gun was put into service in 1942 but was never supplied in enough quantity to replace the 37-mm gun.

Specifications: Type 94 37-mm Antitank Gun	
Caliber	37 mm
Weight	714 lb
Length	9.5 ft
Gun tube length	5 ft 9 in
Projectile weight	1.52 lb
Width	3 ft 11 in
Breech	sliding block
Elevation	–10° to +25°
Traverse	60°
Rate of fire	30 rpm
Muzzle velocity	2,300 ft/sec
Max range	4,900 yd
Sight	straight telescope

Japanese Type 94 37-mm antitank gun. (Wikimedia Commons)

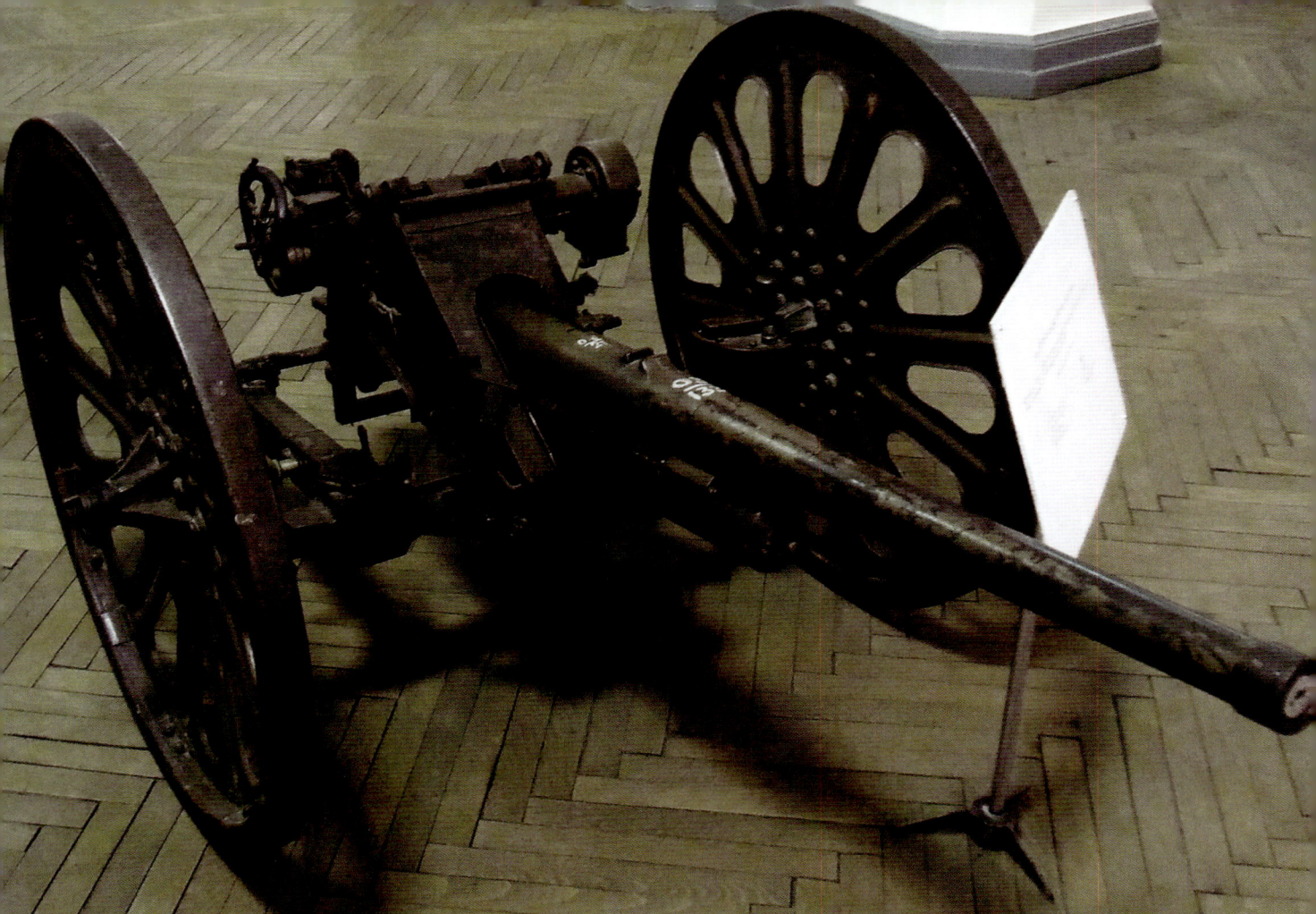

Japanese Type 94 37-mm antitank gun. (Wikimedia Commons)

Polish Armata przeciwpancerna 37-mm wz. 36 Bofors in World War II

Poland purchased a license to manufacture the Swedish Bofors antitank gun, designating it as Armata przeciwpancerna 37-mm wz. 36 Bofors. The Stowarzyszenie Mechanikyw Polskich z Ameryki (SMPzA) Company near Warsaw manufactured the Polish version of the gun, which was the same as the Swedish version except for the tires. The wheels were the main difference between the Polish and Swedish guns. The Polish had approximately 1,200 Bofors guns at the beginning of World War II. During the Polish campaign, the Germans lost approximately 800 armored vehicles, many at the hands of the Armata przeciwpancerna 37-mm wz. 36 Bofors.

Polish Armata przeciwpancerna 37-mm wz. 36 Bofors. (Wikimedia Commons)

Specifications: Armata przeciwpancerna 37-mm wz. 36 Bofors

Caliber	1.45 in
Weight	816 lb
Length	10 ft
Barrel length	62.25 in
Width	3 ft 7 in
Height	3 ft 5 in
Cartridge	1.45 in x 9.80 in
Breech	dropping block
Elevation	-10° to +25°
Traverse	50°
Rate of fire	12 rpm
Muzzle velocity	2,652-2.854 ft/sec
Max range	7,108 yd

Russian 37-mm M1939 (61-K) Antiaircraft Gun and M1930 (1-K) Antitank Gun

37-mm antitank gun model 1930 (1-K) (Wikimedia Commons)

In January 1938, the Artillery Factory Number 8 in Sverdlovsk developed a 37-mm antiaircraft gun (61-K) that was a downsized copy of the Bofors 40-mm AA gun. Trials of the 61-K occurred in October 1938. The performance was similar to the Bofors so the 61-K was adopted in 1940.

The 37-mm antitank gun (1-K) was the first developed and used by the Soviets. It was compact and easily operated by the crew. It suffered from poor manufacturing techniques and ammunition performance because of inferior propellant. By 1941, the 1-K was effective against lightly armored vehicles and side shots on more heavily armored vehicles. The 1-K could also fire German ammunition which enhanced the gun's utility and performance. The gun had a maximum penetration of 1.65 inches perpendicular to rolled homogenous plate at 109 yards and 2.40 inches penetration perpendicular to hardened carbon steel plate. The performance was inferior to the German Pak 36.

37-mm automatic air defense gun M1939 (61-K). (Wikimedia Commons)

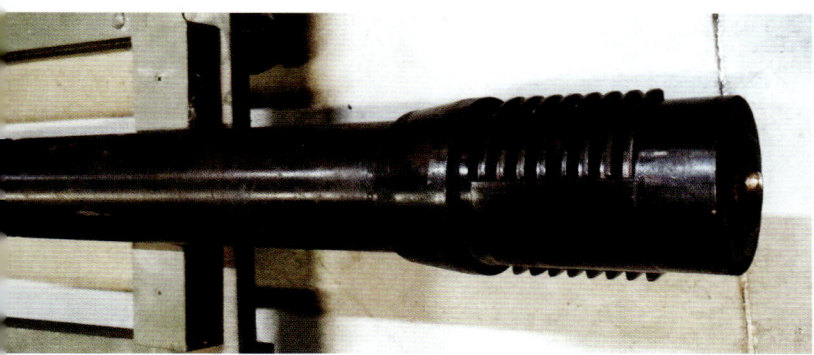

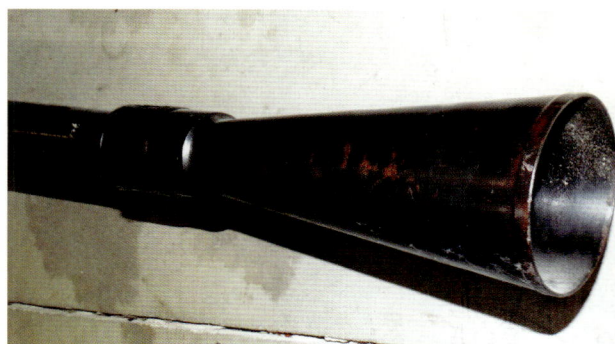

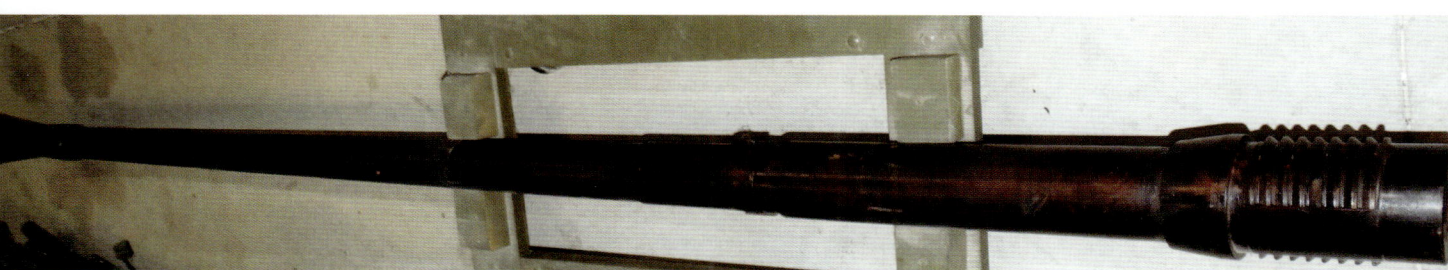

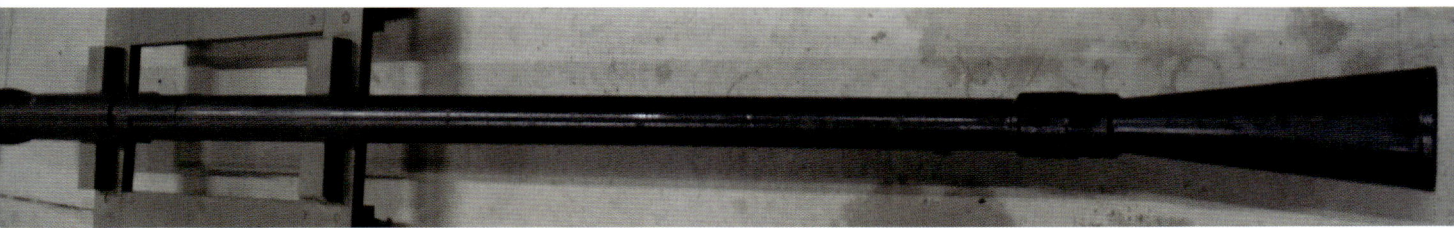

(Top left) Russian 37-mm antiaircraft Gun M1939 (61-K) breech end. (Roberts Collection)

(Top right) Russian 37-mm Gun M1939 (61-K) muzzle end showing flash suppressor. (Roberts Collection)

(Middle) Russian 37-mm Gun M1939 (61-K) barrel, breech end. (Roberts Collection)

(Bottom) Russian 37-mm Gun M1939 (61-K) barrel, muzzle end. (Roberts Collection)

Specifications: Model (1-K) Antitank Gun

Caliber	1.45 in
Gun tube length	5 ft 5 in
Breech	sliding block
Elevation	-8° to +25°
Traverse	60°
Rate of fire	10–15 rpm
Muzzle velocity	2,600–2,800 ft/sec

Czechoslovakian 37-mm KPUV vz. 37 Antitank Gun (Wikimedia Commons)

In Profile:
Czechoslovakian 37-mm KPUV vz. 37 Antitank Gun

(Roberts Collection)

The 37-mm KPUV vz. 37 was an antitank gun manufactured by the Skoda Works in Czechoslovakia and was used in World War II. The Skoda Works manufactured other armaments including tanks. The gun was originally designated 37-mm A3 and redesignated 3,7-cm KPUV vz. 34 (Kanon Proti Utocne Vozbe, literally translated as "cannon against the attack car"). The gun was designed for use by the Czechoslovakian Army. A newer version by Skoda, designated 37-mm A4 tested favorably and was also adopted by the Czechoslovakian Army under the designation 3,7-cm KPUV vz. 37. The A4 guns were sold to Yugoslavia, Bulgaria, Hungary, and other countries. After the Germans occupied Czechoslovakia in 1939, the guns were captured and redesignated 3,7-cm Pak 37(t). The guns captured from the Yugoslavs were redesignated 3,7-cm Pak 156(j). The carriage was equipped with wooden spoked wheels or pneumatic tires. The Germans discovered that the A4 gun was superior to the Pak 36 and hence the adoption and redesignation 3,7-cm Pak 37(t).

American M3A1 37-mm gun on M4 carriage showing muzzle gas deflector. (Roberts Collection)

U.S. 37-mm Gun in World War II

As World War II approached the U.S. was behind in many types of weapons including antitank artillery. Army units at the time were using .50-caliber machine guns but needed something heavier for taking out tanks.

In 1936, the Spanish Civil War broke out and the Germans became involved on the side of the Nationalists headed by General Francisco Franco who opposed the Republican government. The Germans sent in armored units and other military equipment to help the Nationalists, including the 37-mm Pak 35/36 which was effective against tanks at that time. This involvement was also an opportunity to utilize and test the latest German military equipment as well as secure an ally in the likely war in Europe.

Having received reports on the performance of German antitank guns, the U.S. Ordnance Department procured two German 37-mm antitank guns for evaluation. The intent was to develop an antitank gun that would be utilized by the Infantry branch. Larger-caliber guns were not considered since it was desired that the gun could be handled by the crew. Several other countries had developed a 37-mm gun, suggesting that this caliber was well accepted amongst foreign militaries.

Development testing continued, resulting in the adoption of the T10 gun and T5 gun carriage in December 1938, culminating in the M3 antitank gun. Watervliet Arsenal manufactured the gun while Rock Island Arsenal produced the carriage. Production started in 1940 and lasted until 1943. During the production period, modifications were made to the gun carriage, changing shoulder guard and traverse controls and changing the designation from Carriage M4 to Carriage M4A1 on January 29, 1943.

Not all gun carriages were modified in this way. Another change was a thread cut at the end of the barrel to accept a gas deflector that was supposed to direct propellant gases to the sides and avoid kicking up dust that obscured the gunsight and hindered the rate of fire. Another reason for the muzzle brake was to reduce recoil. A severe safety problem resulted with this muzzle brake when firing canister ammunition. Some of the shot would be deflected by the muzzle brake, causing possible injury to the crew. The muzzle gas deflector was removed and not used in combat. The gun barrel was a single-piece, forged tube with 12-groove rifling, right-hand twist and one turn in 25 projectile diameters. The carriage was a split-trail type with pneumatic tires. The wheels were ordinary automobile wheels with civilian tires. There were wheel segments under the axle that could be lowered to increase stability when firing. All controls, elevation, traverse, and gunsight were set on the left side of the carriage so that one person could operate the gun.

The 37-mm M3 towed antitank gun was the first to enter the service, usually towed by Jeeps, weapons carriers or halftracks. The U.S. Army, airborne, infantry, mountain and armored divisions were issued the M3 towed antitank gun in large numbers. M3 towed antitank guns were sent to the Nationalist Chinese, Britain, France and the Soviet Union. The M3 first saw combat during the defense of the Philippines. The gun was effective against Japanese tanks which were lightly armored. The Japanese apparently did not value the protection often needed for crews of equipment as exemplified by Japanese aircraft that did not have protective armor for the pilots. The U.S. Marines utilized the M3 effectively against Japanese bunkers and other strongpoints. The war in Europe was a different situation. German armor development progressed at a rapid pace and the M3 antitank guns became ineffective around 1943 and were reassigned to the South Pacific. In Europe the gun was replaced by the British 6-pounder (57-mm Gun M1). A limited number of M4/M9 auto cannons were used on American naval vessels such as PT boats. Mounted on the foredeck of a PT boat, the P-39 aircraft version of the gun was utilized in attacks against Japanese barges and proved effective.

Year	1940	1941	1942	1943	Total
U.S. 37-mm antitank gun production	340	2,252	11,812	4,298	18,702

(Wikimedia Commons)

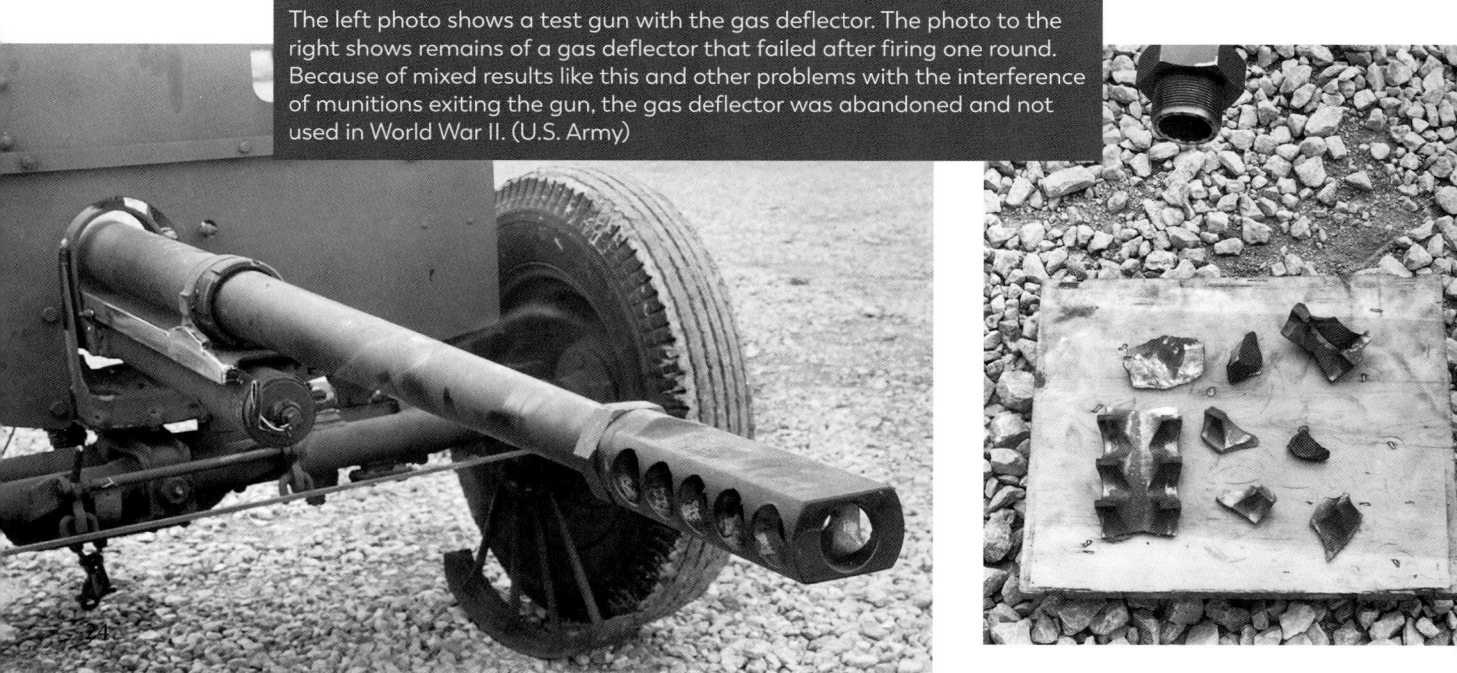

The left photo shows a test gun with the gas deflector. The photo to the right shows remains of a gas deflector that failed after firing one round. Because of mixed results like this and other problems with the interference of munitions exiting the gun, the gas deflector was abandoned and not used in World War II. (U.S. Army)

Side view
of 37-mm
antitank gun
M3 and crew.
(FM 23-70)

Rear view of 37-mm
antitank gun M3 and
crew. (FM 23-70)

| 2 U.S. 37-mm Gun Design

There were several different 37-mm guns used by the American forces in World War II. Each gun, with the exception of the M1916, was equipped with a variety of ammunition for usage against personnel, general targets, armor plate and target practice.

These were:

- 37-mm Gun M1916 which was left over from World War I but was used for training and as a sub-caliber gun for aiming larger guns.

- 37-mm Gun M1A2 mounted on a towed carriage or M15A1 halftrack and used for antiaircraft defense and as an antitank gun.

- 37-mm Gun M3 and M3A1 mounted on a towed carriage, used as an antitank gun and for attacking infantry strongpoints.

- 37-mm Automatic Gun M4, M9, and M10 used in aircraft applications (P-39 Airacobra and P-63 Kingcobra).

- 37-mm Guns M5 and M6 mounted in tanks.

- 37-mm T32 and T33 infantry guns.

37-mm Gun M1916

The 37-mm Gun M1916 was obsolete at the beginning of World War II. It was used on larger guns as a sub-caliber training device. For training purposes, the sub-caliber gun had similar characteristics as that of the larger gun, thereby eliminating the expense of the larger ammunition and the wear and tear on the larger gun. The sub-caliber gun would fire, and the performance of the gun crew could be evaluated based on the 37-mm gun performance.

Specifications: 37-mm M1916	
Caliber	1.45 in
Weight	238 lb
Barrel length	2 ft 5 in
Cartridge size	1.45 in x 3.7 in
Elevation	-8° to +17°
Traverse	35°
Rate of fire	25 rpm
Muzzle velocity	1,200 ft/sec
Maximum range	2,600 yd
Effective range	1,600 yd

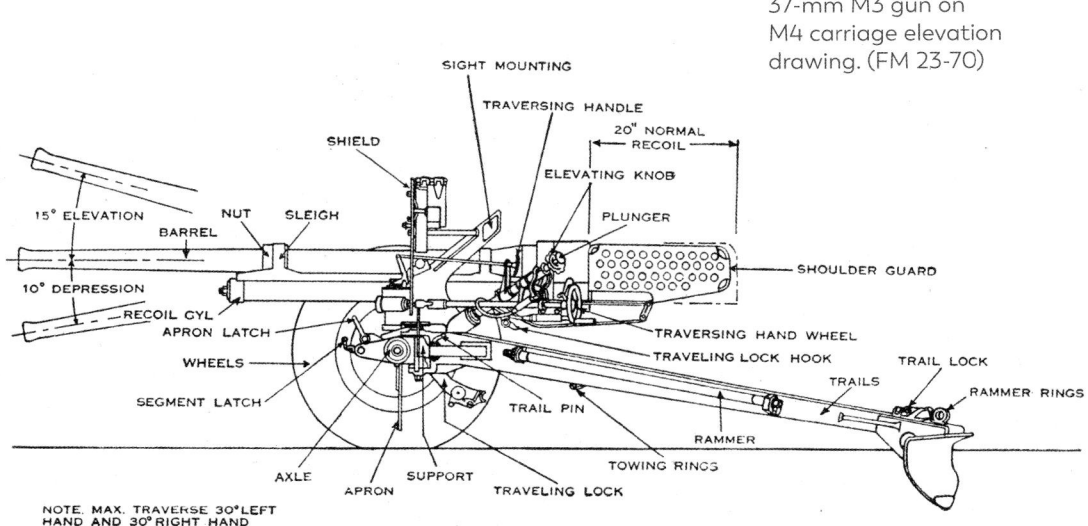

M1916 sub-caliber gun mounted on a 155-mm howitzer. (U.S. Army)

37-mm M3 gun on M4 carriage elevation drawing. (FM 23-70)

SIGHT MOUNTING

TRAVERSING HANDLE

20" NORMAL RECOIL

SHIELD

ELEVATING KNOB

15° ELEVATION

PLUNGER

NUT   SLEIGH

BARREL

SHOULDER GUARD

10° DEPRESSION

RECOIL CYL

TRAVERSING HAND WHEEL

APRON LATCH

TRAVELING LOCK HOOK

WHEELS

TRAIL LOCK

SEGMENT LATCH

TRAILS

RAMMER RINGS

TRAIL PIN

AXLE   APRON   SUPPORT   TRAVELING LOCK

RAMMER

TOWING RINGS

NOTE. MAX. TRAVERSE 30° LEFT HAND AND 30° RIGHT HAND

37-mm M3 and M3A1 on M4 and M4A1 Carriages

The 37-mm antitank gun, M3, was a single-shot, flat-trajectory field gun utilizing a dropping block (vertical sliding block) type breechblock. It fired projectiles weighing approximately 2 lb. It was mounted on an M4 carriage of split trails with pneumatic tires. The gun was designed such that one man could aim the weapon using elevation and traversing controls. One man could load the weapon, fire it, and eject the spent shell. The gun was also designed to be towed by military vehicles via a lunette using a pintle hitch. The data in the textbox characterizes.

Specifications: 37-mm M3 Towed Gun

Weight of gun tube	191 lb
Weight of gun and carriage	912 lb
Height of gun bore above ground	25.75 in
Length of carriage from muzzle to lunette	154.5 in
Overall width	63.5 in
Width of tread, centerline of wheels	56 in
Trail spread maximum included angle	60°
Elevation	-10° to +15°
Max traverse right or left	30°
Recoil mechanism: hydro-spring normal recoil	20 in
Tires 6 ply pressure 10 psi	6.00x16

The gun tube was a one-piece forging with a rifled bore and a large threaded end to accept the breech. Yield strength of the steel was approximately 95,000 psi. (The yield strength of a metal is that stress level that, when exceeded, deforms the metal and it does not return to its original shape after the stress is reduced. Consequently, gun designers keep the stress in the barrel way below this figure to avoid damage to the gun.) There were two bearings, one at the middle of the gun tube to support and align the barrel and the other to secure the sleigh during travel.

The gun was mounted on either a carriage M4 or M4A1. The difference between the two mostly dealt with the firing plunger versus firing handle and the shape of the shoulder guard. The pneumatic tires were typically civilian truck type and not the typical military tread since traction was not an issue and the tires were easily obtained. Wheel segments were mounted near each wheel to stabilize the gun when firing. Ramrods were attached to each of the trails for cleaning out the gun tube. The ramrods were in two pieces with threaded ends so they could be connected to form a pole long enough for the gun tube. There was a gear case mounted below the gun for elevation adjustment by a hand wheel. There was another gear case just ahead of the shield on the left that controlled the traversing via a knob. A quick disconnect puller was available to disengage the traversing gear box so one could rapidly traverse the gun from one side to the other. Shown on page 31 is the recoil mechanism. This is a hydro-spring design where both spring action and hydraulic fluid pressure absorb the recoil energy of the gun. After the gun is fired, the center recoil mechanism returns the gun into battery (normal firing position), ready for the next shot.

37-mm M1916 sub-caliber gun mounted
on a 75-mm howitzer. (U.S. Army)

When the M3 37-mm gun is fired, the pressure in the gun rises quickly as the projectile starts to move toward the muzzle. After about a quarter way down the gun tube, the pressure starts to drop as the volume in the tube increases and the projectile accelerates. The velocity of the projectile increases and reaches a maximum as it exits the gun tube. The tangential resistance in psi is the stress in the tube steel as the round travels by. As shown in the drawing on the opposite page, the maximum tangential stress occurs at the firing chamber at the breech end of the gun tube and decreases along the length of the gun tube. The yield strength of the gun tube steel is 95,000 psi.

Interior Ballistics of the 37-mm M3 Gun

Interior ballistics is the study of parameters that affect how a projectile travels along the length of the gun tube as a result of propellant burn and is then directed toward a target. As the propellant charge burns inside the cartridge case, gas pressure rises significantly, causing the projectile to move toward the muzzle. As this happens, the drive band (rotating band) on the projectile contacts the rifling grooves which deform the band but also cause the projectile to rotate. This is called engraving and causes the projectile to spin. The spin increases the accuracy of the shot, thereby reducing pitch and yaw gyrations that deflect the projectile from its expected path. The M3 gun rifling was 12 grooves with right-hand twist when looking from the breech end to the muzzle end. The pitch was one turn in 25 calibers or one turn in approximately 36 inches. The caliber is the diameter of the gun tube between lands at opposite ends.

The rifling in a gun tube has grooves and lands, the lands being the raised portion of the gun tube and the grooves being the spaces between the lands. The 37-mm M3 gun is often referred to as L/56.6. L is the length of the gun from the rear face of the breech to the muzzle and is divided by the bore diameter. Consequently 56.6 = 82.5 inches divided by 37-mm (1.456 inches).

The 37-mm M3 gun tube or barrel was constructed of steel with a yield strength of 95,000 psi. The yield strength in steel is a property whereby if the loading from firing a shot results in a stress higher than 95,000 psi, the steel will deform and not return to its original shape. Therefore, the propellant charge in the 37-mm cartridge must be sufficient to generate enough pressure to expel the projectile at the required velocity but not be strong enough to damage the breech end of the gun. In the curve on the facing page, we see that maximum pressure allowed in the gun tube based on material properties and gun dimensions.

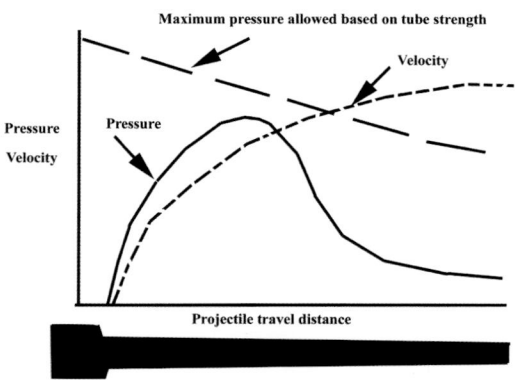

Pressure-travel and velocity-travel curves (Roberts Collection)

The maximum allowable pressure decreases as a projectile travels along the tube because the tube outer diameter decreases. However, as the projectile travels along the tube, the pressure starts to drop because of the increased volume behind the projectile and propellant burn out. Consequently, it is typical of large guns that the maximum pressure is at the breech end, and that the pressure drops off significantly at the muzzle end, hence the thinner wall of the 37-mm gun tube at the muzzle end and the tapered shape.

As shown in the drawing to the left, the projectile overcomes the friction of engraving the grooves as the pressure rises, and the velocity increases. The drawing on the facing page shows detailed pressure,

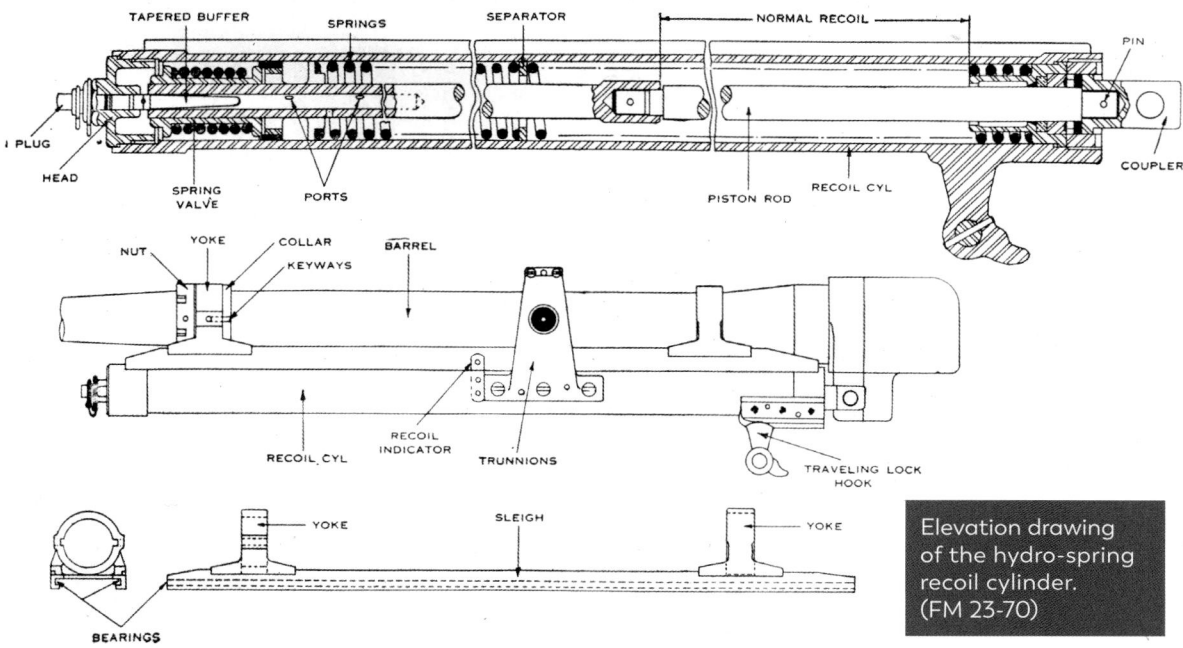

Elevation drawing
of the hydro-spring
recoil cylinder.
(FM 23-70)

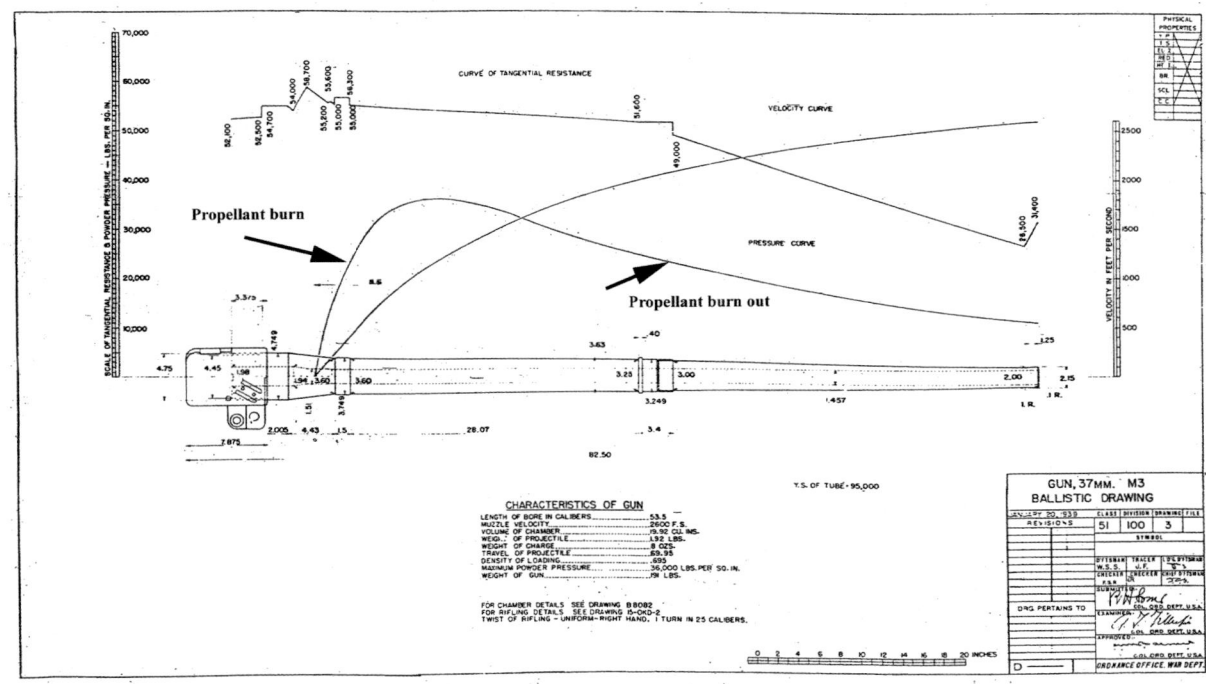

Gun, 37-mm M3. Ballistic drawing showing propellant pressure distribution, velocity curve, and tangential resistance for a typical cartridge. (Roberts Collection)

velocity, and friction curves for the 37-mm M3 gun firing a projectile with a muzzle exit velocity of 2,600 ft/sec. The peak propellant burning pressure against the projectile is approximately 36,000 psi. The tangential resistance reaching a maximum of 58,700 psi is essentially the stress in the gun tube material as a result of the propellant pressure buildup. It is sufficiently below the 95,000 psi yield strength of the material, yielding an adequate design.

There were a variety of cartridges used in the 37-mm M3 gun depending on their intended use. Their important characteristics are weight of the projectile, the amount and type of propellant in the cartridge case, and the design of the drive band (rotating band). The weight of the projectile governs its acceleration. The lighter the projectile, the higher the acceleration rate out of the tube for a given propellant charge. However, a light projectile has less kinetic energy than a heavier one for the same velocity, which makes a difference on impact with a target (terminal ballistics).

The propellant grain characteristics and the amount of propellant play a major role in the shaping of the pressure–travel curve. Grain size and shape control the rate of burning of the propellant. Grains may be small cylindrical pellets, small spherical shapes, small strips, and perforated shapes. These different shapes have various surface areas, which govern the burn rate of the propellant. Burn rate must be controlled to avoid a rapid pressure spike that can damage the breech end of the gun. However, the burn rate must be able to drive the projectile out of the gun tube with sufficient pressure at the muzzle end. Surface area of the propellant grain has a significant influence on burn rate. The higher the surface area, the higher the burn rate.

Finally, the drive band for spin-stabilized projectiles is very important. The band is slightly larger than the tube diameter and must limit leakage of propellant gases forward of the projectile. It must be a soft metal so that it deforms and forms grooves, which is called engraving. There is a high initial resistance of the projectile to movement, but when this resistance is overcome by the propellant pressure, it moves along the gun tube forming a seal and a means of rotating the projectile.

The pressure–travel curve in the graph on the previous page shows the pressure that will give a muzzle velocity of 2,600 ft/sec. This is below the capability of the gun to deliver a projectile at a higher speed, which is a compromise. Any higher pressure–travel curve may require additional material at the breech end of the gun, reduce the life accuracy of the gun as a result of increased erosion, and cause a brighter flash as the projectile leaves the muzzle.

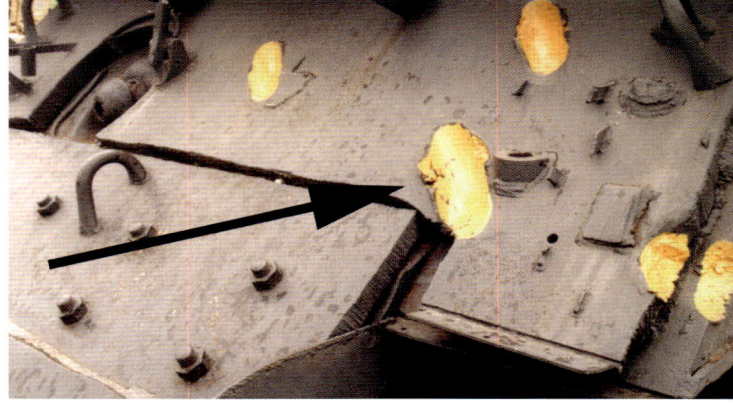

Ricochet impacts from a large-caliber weapon on the glacis plate of a Sherman tank. The projectile did not penetrate the armor but may have caused spalling inside the fighting compartment. (Roberts Collection)

Exterior Ballistics

Exterior ballistics is the study of parameters that affect how a projectile travels through the air on the way to its target. Like many other high-caliber guns, the 37-mm projectile is subject to several forces during flight. They are gravity, aerodynamic drag, supersonic wave drag, spin drift, and Coriolis force (the force experienced by an object

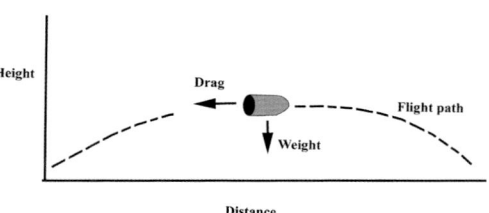

Projectile flight path. (Roberts Collection)

traveling over a rotating object, the earth.) Other factors affecting the accuracy of a projectile are altitude, wind, temperature, and humidity. Gravity and aerodynamic drag are probably the most significant factors in affecting the flight of a projectile. For a lower-velocity 37-mm projectile, drop can be a significant aspect of aiming the gun. For the higher velocity shells, this is less of a factor.

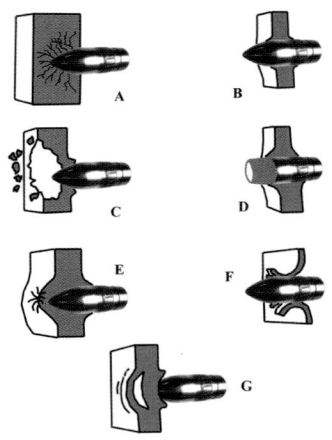

Terminal ballistic penetration modes. (Roberts Collection)

Terminal Ballistics

Terminal ballistics is the study of the interaction of a projectile with an object, the target. Typical penetration modes of a solid 37-mm projectile into steel are shown on the left. (A) is brittle fracture of target metal. (B) is ductile penetration into a target without brittle fracture. (C) is a radial fracture of the target material. (D) is plugging where the projectile is flattened and shears out a plug of target metal. (E) is fragmentation. (F) is a ductile deformation in a soft target material. (G) is spalling where the initial compression wave reflects off the back of the plate causing chunks of metal to separate on the opposite side of impact. This often caused injuries to the crew inside the armored vehicle. Typically, projectile impacts are multimodal, meaning that the failure mode of the target is a result of several different modes. The following images present actual examples of projectile impacts on armor.

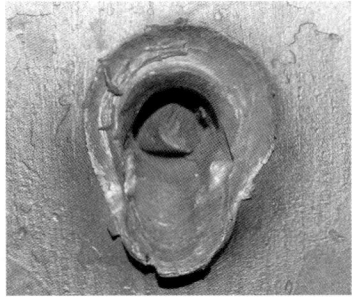
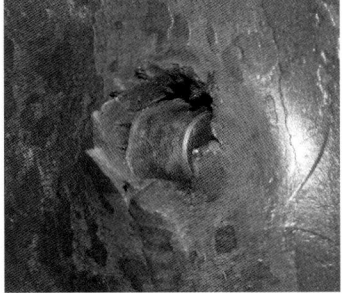

(Left) Complete penetration from a large-caliber weapon on a Sherman tank frontal armor. There are also some small-caliber impacts that did not penetrate the armor. (C. Roberts, *Armored Strike Force: The Photo History of the 70th Tank Battalion in World War II*); (Middle) Entry crater of a 14.5-mm projectile from a Russian PTRS41 antitank rifle into 1-inch-thick homogeneous steel plate. (Roberts Collection); (Right) Exit hole of a 14.5-mm projectile from a Russian PTRS41 antitank rifle into 1-inch-thick homogeneous steel plate. (Roberts Collection)

37-mm Gun M3 Gun Components

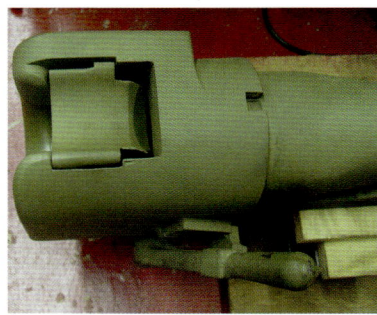

(Clockwise from top left):

Top view of the breech of an M3A1 37-mm gun in the closed position. (Roberts Collection)

M3 gun tube on the left, M3A1 gun tube (without breech) on the right. The difference is the threaded muzzle end on the M3A1. (Roberts Collection)

Muzzle end of the 37-mm M3A1 showing a threaded end to accept a gas deflector. The gas deflector proved to be a problem when firing canister ammunition, so it was removed and not used, as canister balls would interfere with the gas deflector. It is interesting to note that without the gas deflector, the flared end no longer exists on this model, which probably resulted in quicker crack development. (Roberts Collection)

Muzzle view of 37-mm M3 gun. The outer diameter of the gun tube is highly stressed at the end because of no supporting tube material at the end. The flare at the end of the barrel is to reduce stress as the projectile is expelled, thereby reducing the development of cracks. Cracks in a gun tube are dangerous as they may cause gun tube shrapnel to be released. (Roberts Collection)

View of high-pressure end of an M3A1 (the chamber) showing the threads that mate with the breech. The horizontal groove is clearance for the shell extractor. (Roberts Collection)

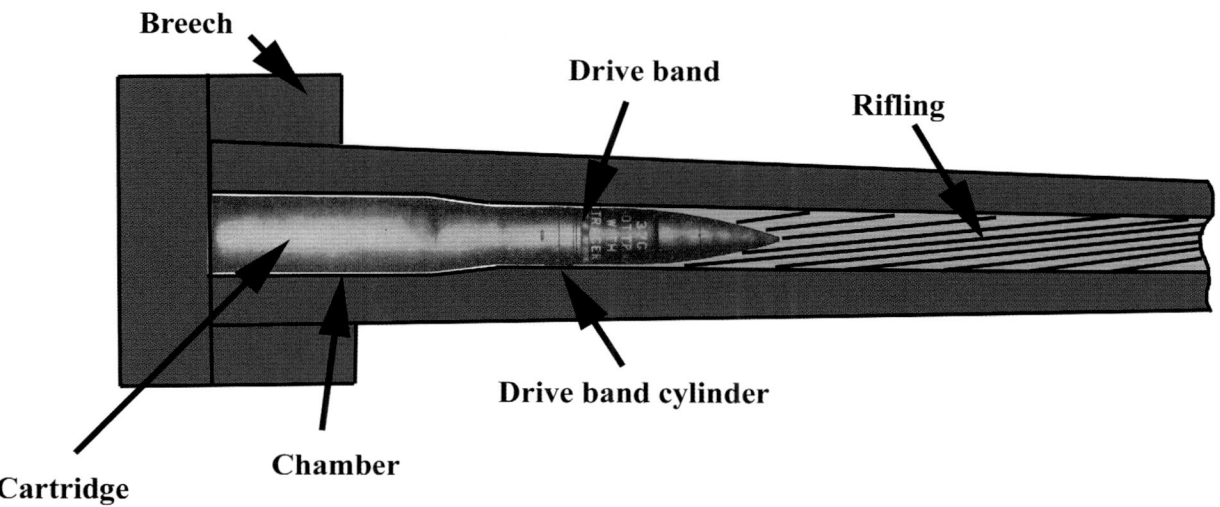

Breech

Drive band

Rifling

Drive band cylinder

Cartridge

Chamber

Major components of the high-pressure end (the chamber) of the gun. (Roberts Collection)

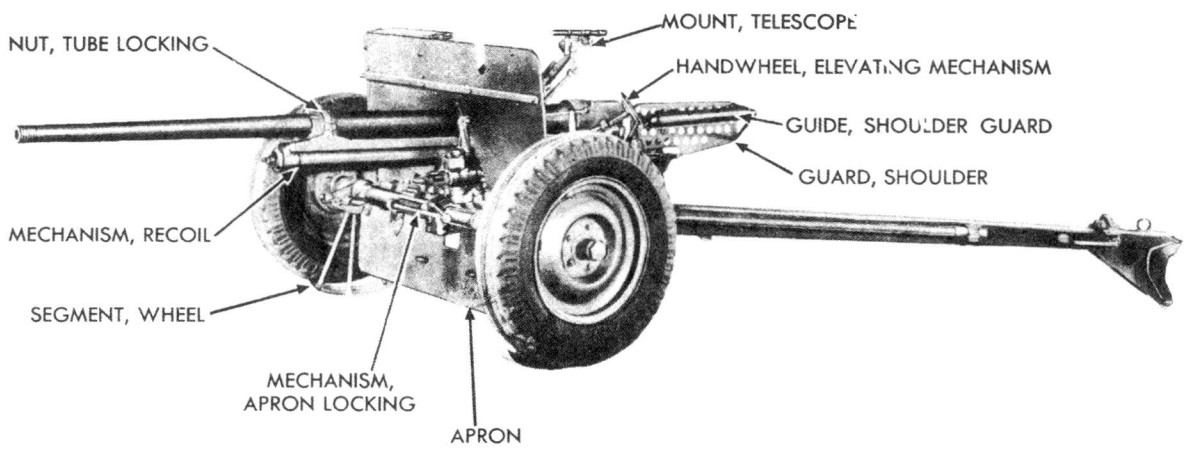

NUT, TUBE LOCKING

MOUNT, TELESCOPE

HANDWHEEL, ELEVATING MECHANISM

GUIDE, SHOULDER GUARD

GUARD, SHOULDER

MECHANISM, RECOIL

SEGMENT, WHEEL

MECHANISM, APRON LOCKING

APRON

37-mm Gun M3A1 on M4A1 carriage in firing position with trails spread. The spade or plow at the end of the trails is designed to dig itself into the ground as a result of the recoil force. The wheel segment is shown in the lowered position to reduce bounce during firing. For movement, it is rotated upward so that the tire contacts the road. This gun is designed for direct fire, i.e., the gun is aimed straight at the target, as opposed to indirect fire where a projectile is lobbed over a hill and strikes a target not observable by the gunner. (FM 23-75)

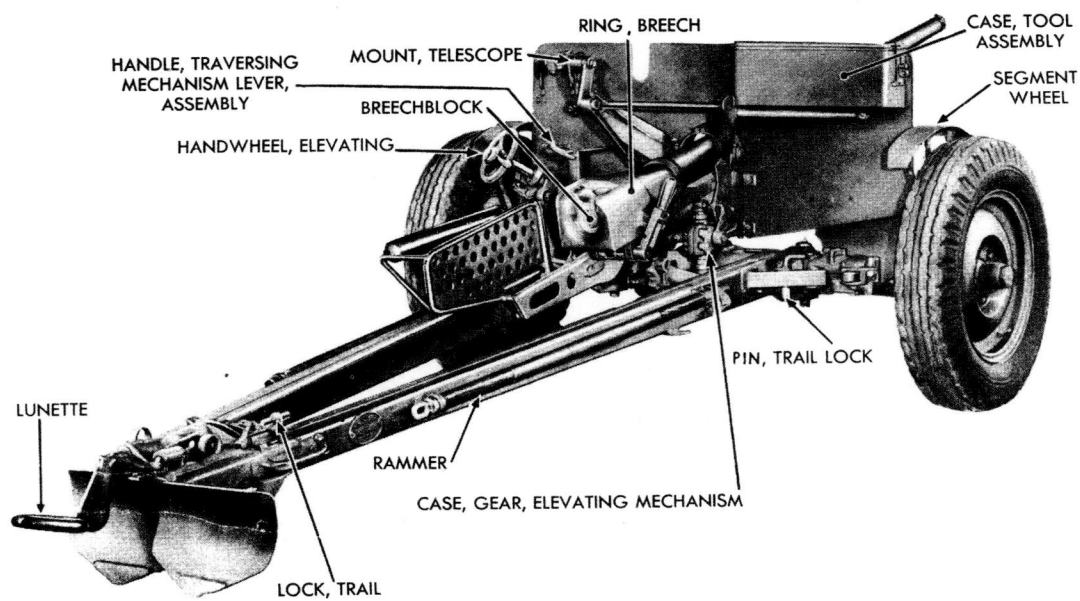

HANDLE, TRAVERSING
MECHANISM LEVER,
ASSEMBLY

MOUNT, TELESCOPE

RING , BREECH

CASE, TOOL
ASSEMBLY

BREECHBLOCK

SEGMENT
WHEEL

HANDWHEEL, ELEVATING

PIN, TRAIL LOCK

LUNETTE

RAMMER

CASE, GEAR, ELEVATING MECHANISM

LOCK, TRAIL

37-mm Gun M3A1 and Carriage M4A1 in traveling configuration with trails folded and secured. The rammer (ramrod) is used to clean the gun tube by wrapping a piece of cloth at the end and pushing it through the gun tube. (FM 23-75)

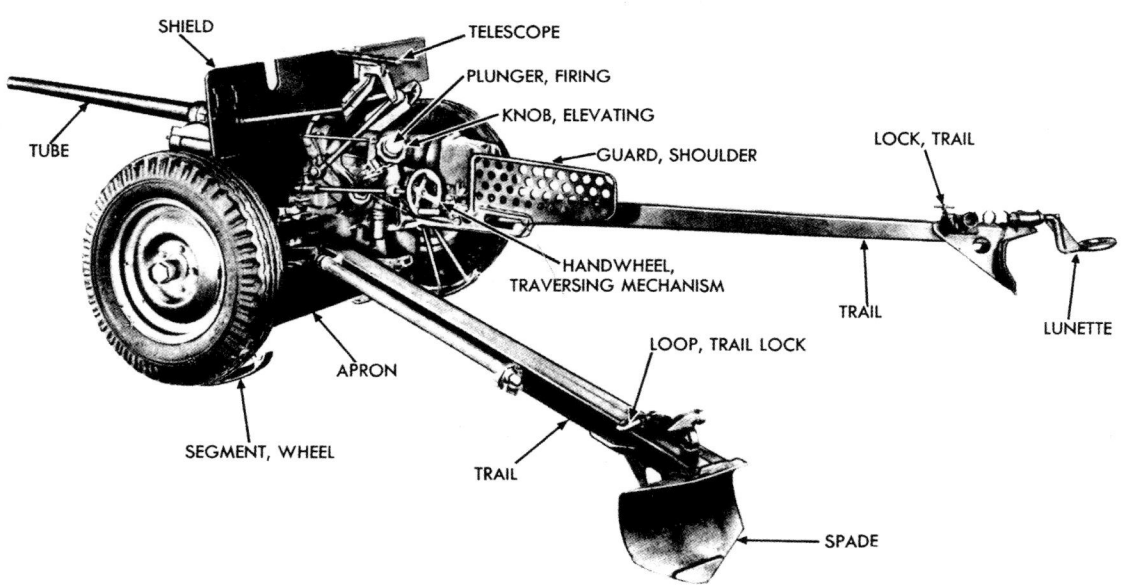

SHIELD

TELESCOPE

PLUNGER, FIRING

KNOB, ELEVATING

LOCK, TRAIL

TUBE

GUARD, SHOULDER

HANDWHEEL,
TRAVERSING MECHANISM

TRAIL

LUNETTE

APRON

LOOP, TRAIL LOCK

SEGMENT, WHEEL

TRAIL

SPADE

37-mm Gun M3 and Carriage M4 in firing position with trails spread. (FM 23-75)

LIGHT, BLACKOUT
TAIL AND
BLACKOUT STOP

CABLE, JUMPER

RA PD 81584

Blackout light mounted on the barrel when the gun carriage is being towed. (FM 23-75)

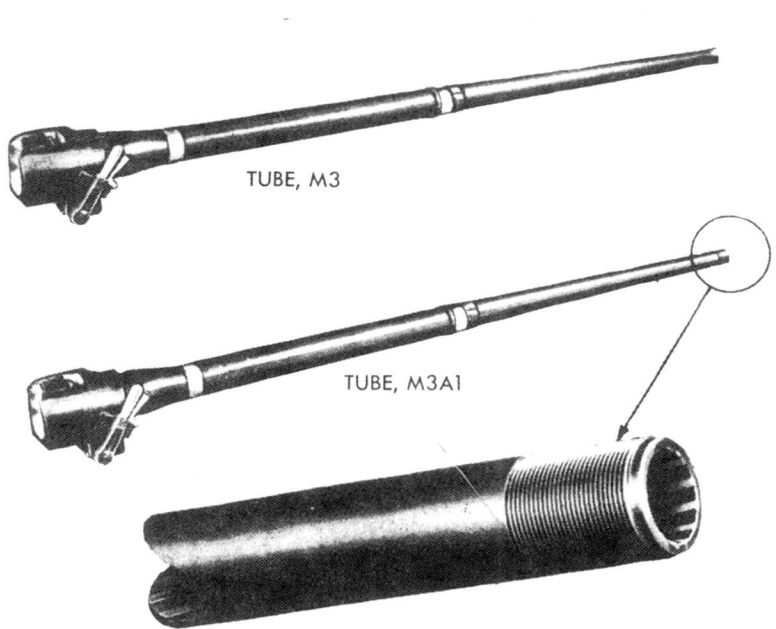

TUBE, M3

TUBE, M3A1

Comparison of gun tubes. The M3 tube does not have a threaded muzzle. The M3A1 has a threaded muzzle to accept a gas deflector. (FM 23-75)

The gas deflector has five holes that deflect exit gases in a manner that reduces the dust evolution in front of the cannon that would obscure aiming. It can be seen that when canister shot (shotgun shell) was used, some of the small balls would get caught in the holes on the side of the gas deflector, causing damage and negating the accuracy of the shot.

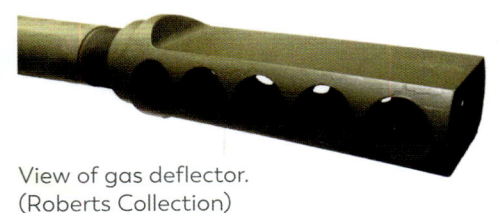

View of gas deflector. (Roberts Collection)

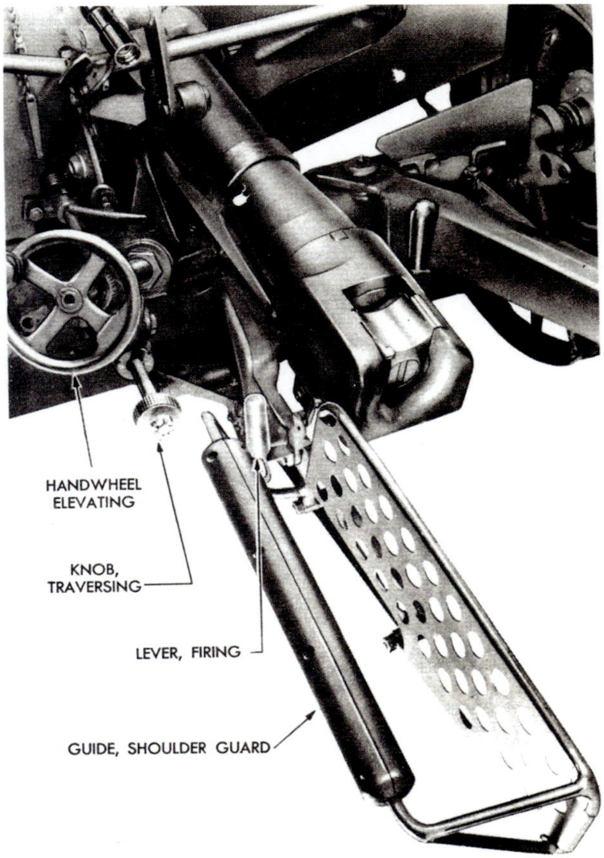

HANDWHEEL ELEVATING

KNOB, TRAVERSING

LEVER, FIRING

GUIDE, SHOULDER GUARD

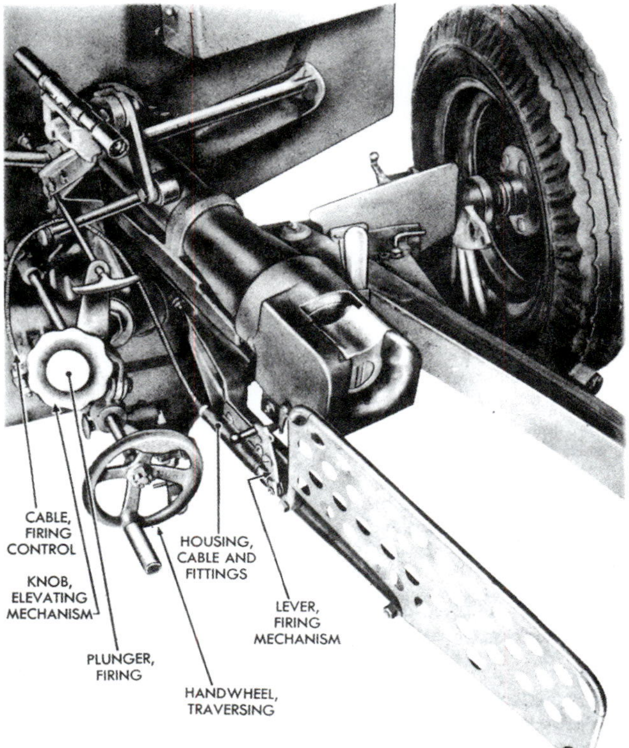

CABLE, FIRING CONTROL

KNOB, ELEVATING MECHANISM

PLUNGER, FIRING

HANDWHEEL, TRAVERSING

HOUSING, CABLE AND FITTINGS

LEVER, FIRING MECHANISM

Carriage M4A1 (left) illustrating the differences between the M4 (right) and M4A1 which are: the position of the hand wheel for elevating the gun, knob for traversing, a firing lever to fire the gun, and a shoulder guard. (FM 23-75)

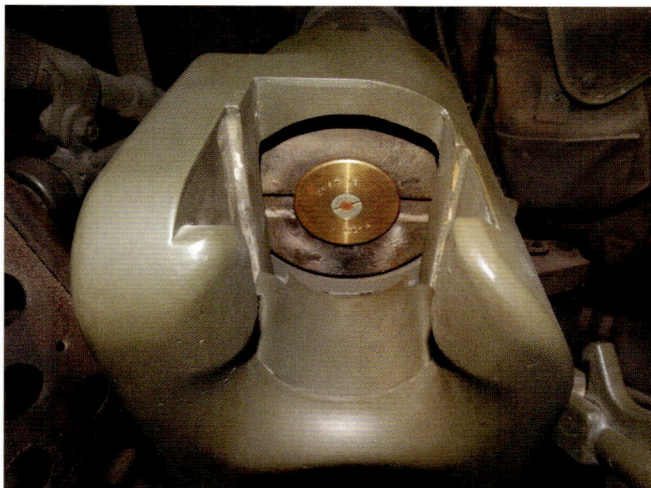

(Clockwise from top left):

View of open breech accepting a 37-mm cartridge. When the handle is pulled back, the shell is extracted. Other breech designs in tanks had an automatic shell ejector system. The spent shell would fall into a storage bag at the back of the gun. (Roberts Collection)

Cartridge in the chamber with dropping block ready to close. The horizontal groove is clearance for the cartridge case extractor. (Roberts Collection)

Carriage M4A1 showing the location of the firing lever, hand wheel for traversing the gun tube, and arm guards. (Roberts Collection)

View of the closed breech on a 37-mm M3 gun. This is a dropping block (vertical sliding block) breech design. The block drops down when the handle is actuated, allowing an opening to insert the cartridge. The thick breechblock is necessary to resist the high pressure at ignition of the propellant. (Roberts Collection)

Experimental conical flash suppresser that was not adopted by the U.S. Army. (U.S. Army)

View of threaded end of the gas deflector. With this mounted on the gun, bore cracks at the muzzle end would probably not develop because of the support from the gas deflector. Without the gas deflector, cracks could easily form. (Roberts Collection)

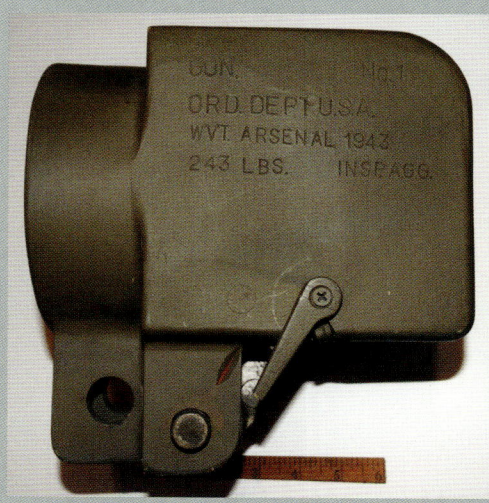

Side view of breech for the 37-mm M3 gun. (Roberts Collection)

A rare photo of a 37-mm M3A1 with the gas deflector attached. This is a media photo. The gas deflector was not used in combat. (U.S. Army)

The breech on an M8 75-mm Pack Howitzer is a horizontal sliding block design. The breechblock slides horizontally when the handle is manipulated, securing the cartridge in the gun. The block is sufficiently thick to withstand the pressure in the gun when fired. The vertical sliding blocks were used with the 37-mm guns and their variants. (Roberts Collection)

There are three types of breech designs: dropping block (vertical sliding block), sliding block (horizontal sliding block), and interrupted thread. Shown below is an example of an interrupted thread breech on a 75-mm recoilless rifle M20. The first photo shows the breech closed, ready to fire. The next photo shows the breechblock screw partially open when the handle is turned clockwise. After the cartridge is inserted into the gun, the breechblock screw is rotated counterclockwise to the fully engaged and locked position. This assures that the cartridge is fully captured and will not back out when fired.

(Top) Breech closed, 75-mm recoilless rifle M20. (Roberts Collection)

(Bottom) Breech partially open, 75-mm recoilless rifle M20. (Roberts Collection)

M6 Sighting Telescope

The M6 telescope used on the 37-mm M3 gun is a straight tube type with no magnification. It has a reticle with a sighting picture. Shown below is the sight picture through the telescope. The dots in the middle represent the elevation of the gun giving various ranges. The higher the elevation, the longer the range. The upper dot is 0 range, the next one down is 600-yard range, the next is the intersection of the circle and a line which is the 1200-yard range, and finally the lowest dot is the 1,500-yard range. The 5-mil lead is the circle, the 10-mil lead is the right or left dot and the 20-mil lead is the far right or left dot. A mil is a milliradian or about 0.06°. The parallelogram M19 telescope mount, shown on the following page, moves with the gun tube and remains aligned with the tube. The M6 sighting telescope is used for all firings of the 37-mm M3 gun on the M4 carriage.

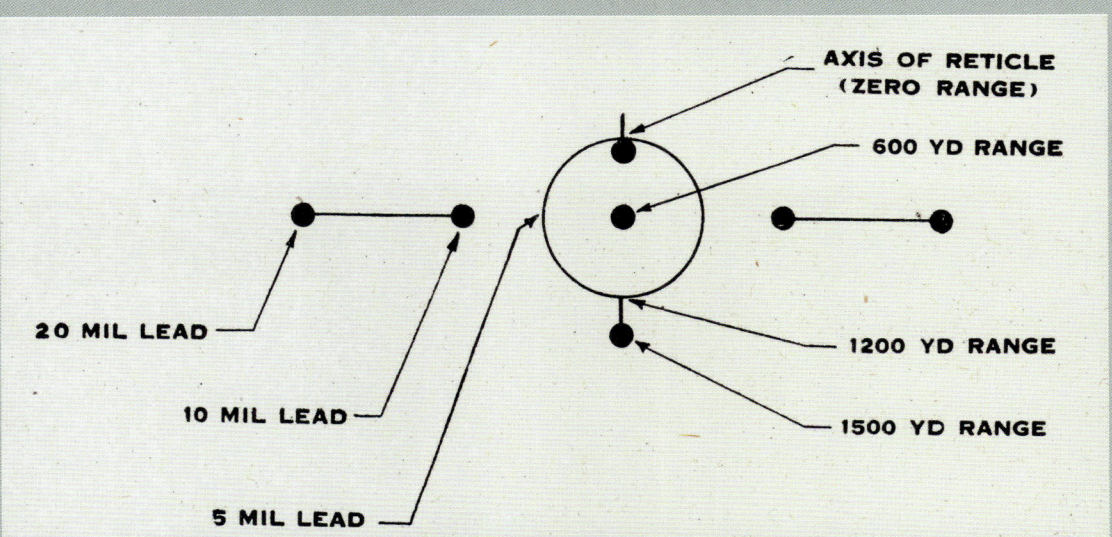

(Top) M6 sighting telescope. (Roberts Collection)

(Bottom) Sight picture through the M6 telescope. (FM 23-70)

Bore Sighting

Bore sighting is a method of accurately aiming a 37-mm gun, by looking through a peep hole at the breech end and sighting along the barrel to cross-hairs at the muzzle end. The primary usage of bore sighting was to align the gun barrel with the gun sight. It was also used against small targets that required high accuracy, such as a Japanese aperture in a pill box.

(Left) Muzzle end of a 37-mm gun M3 showing four grooves used for bore sighting the gun (grooves are at 12, 3, 6, and 9 o'clock). Bore sighting is an accurate way to aim a gun at a very small target. (Roberts Collection)

(Below) M6 sighting telescope on an M19 telescope mount. (Roberts Collection)

In the Far East, the Japanese machine-gun positions would be in a rock cliff with only a small opening for the machine gun. Hitting the rock cliff with cannon fire was not usually effective. However, bore sighting the small Japanese machine-gun hole was effective, resulting in knocking out Japanese positions with a 37-mm high-explosive round that entered the hole. String is placed in these grooves to form a cross that indicates the center of the bore. The threaded setup is shown in the drawing and photo below.

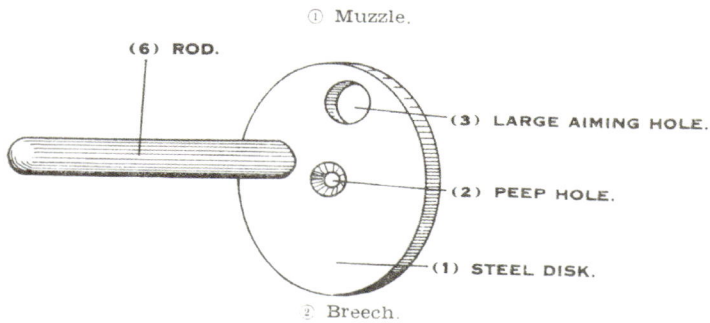

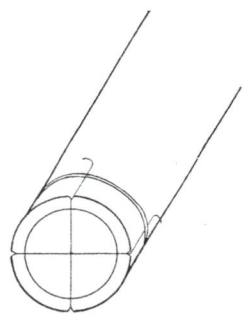

(Top left) Breech-end fixture used for bore sighting. (6) is a rod to hold the device at the breech hole. (1) is the disc that fits the hole at the breech end. The peephole (2) and large aiming hole (3) are used for sighting the cross hairs at the muzzle. (C. Roberts, *U.S. Airborne Tanks 1939–1945*)

(Top right) Drawing of string at the muzzle end of the gun. (Roberts Collection)

(Left) View of string attached to the muzzle end for bore sighting on a 37-mm gun M6, tank mount. The gun could then be fired and the thread had to be reinstalled for the next bore sighting. The crosshairs are held in place by additional wraps of thread around the muzzle end. (Roberts Collection)

37-mm Gun, Tank M5

The 37-mm tank gun M5 is a flat trajectory weapon of the field type very much like the M3 37-mm gun on a towed carriage. It was used on several tanks. It is a single shot gun, has a dropping block breech, and fires a projectile weighing approximately 2 lb. The tank mounts allow manipulation of the gun by hand wheels and turret traversing mechanism. Data on this gun is in the textbox.

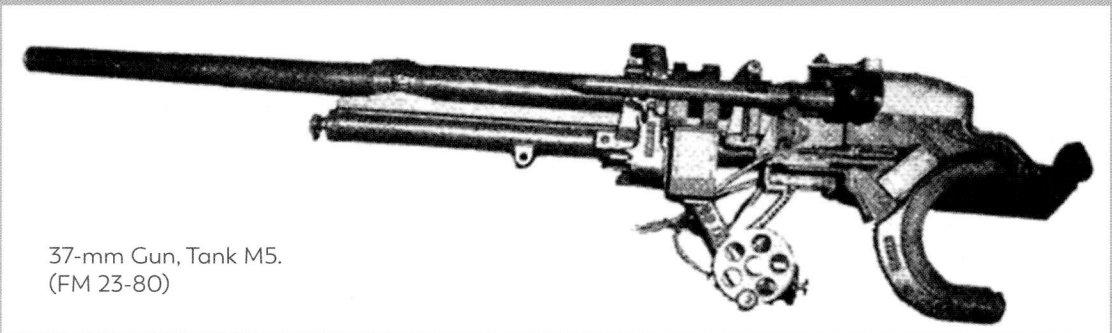

37-mm Gun, Tank M5.
(FM 23-80)

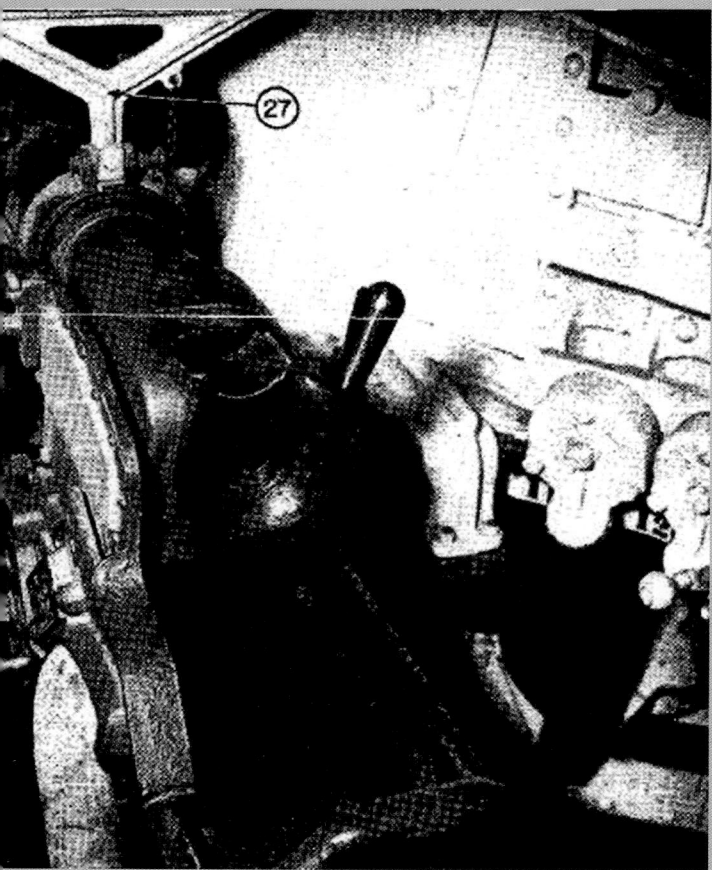

Specifications: 37-mm Gun, Tank M5

Weight.......................................700 lb

Weight of gun tube ....185 lb

Length of gun tube ....73 in

Length of recoil...............................6–8 in

Ammunition........................Canister, AP M51, HE Mk II, HE M63

(Left) 37-mm gun M5 showing manually operated breech. (FM 23-80)

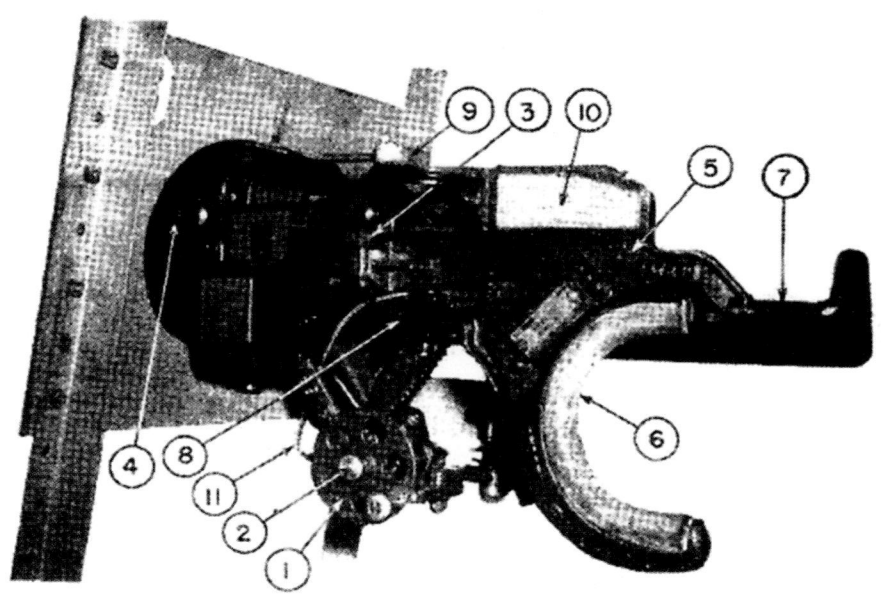

Left view of breech end of 37-mm M5 gun. (FM 23-80)

1. Elevating handwheel
2. 37-mm trigger actuator plunger
3. Rear sight bracket
4. Front sight bracket
5. Shoulder guard
6. Shoulder rest
7. Recoil guard
8. Hand bracket (left)
9. Headrest (forward)
10. Headrest (right side)
11. 37-mm trigger actuating cable

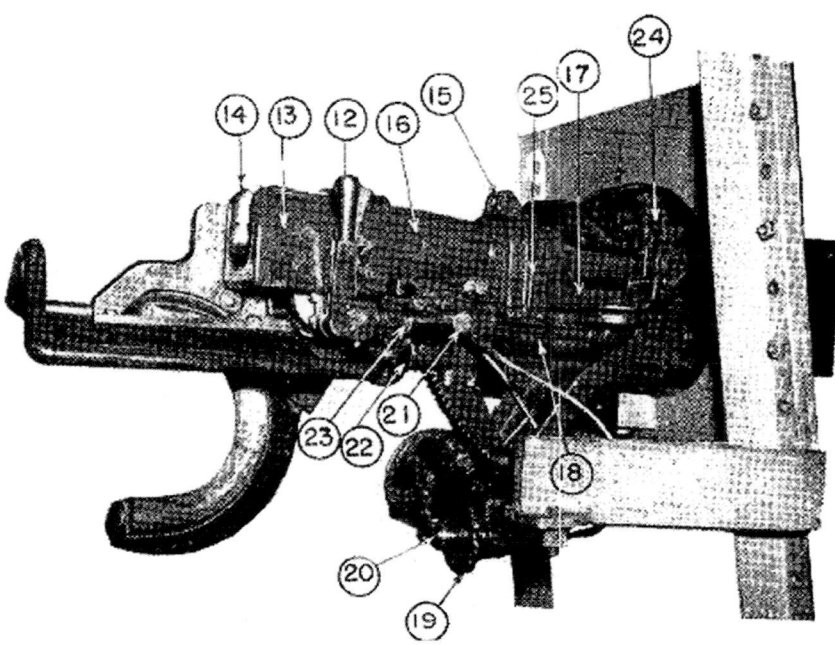

Right view of breech end of 37-mm M5 gun. (FM 23-80)

12. Operating handle
13. Breech ring
14. Cocking lever
15. Traveling lock hook
16. Barrel
17. Cradle
18. Recoil cylinder
19. Traversing handwheel
20. Caliber .30 machine gun trigger actuator plunger
21. Hand bracket (right)
22. 37-mm trigger bar actuator
23. Caliber .30 machine gun trigger actuator
24. Trunnion
25. Yoke

37-mm Gun, Tank M6

The 37-mm tank gun M6 was essentially the same as the 37-mm gun M5 with the following exceptions. The M6 gun had a longer gun tube and the breechblock was operated automatically, therefore no lever was provided to manually eject the spent shell. The M6 also had a gun-leveling feature that is discussed following. See textbox for basic data.

Specifications: 37-mm Gun, Tank M6

Total weight of gun .... 700 lb

Length of gun tube .... 78 in

Length of recoil..6–8 in

Gun elevation .... -10° to +20°

The two drawings are of the operation of the gun stabilizer system for the M6 gun. In the first drawing, the tank tilts upward such that the gun points upward. In order to keep the gun tube level, the hydraulic piston attached to the gun must drop the gun tube to compensate. When the gun tube tilts upward, the gyroscope attached to it precesses (moves), causing the arm attached to it to press against several resistor tabs. If two tabs touch, it is similar to a switch, and current flows to control valves that cause the gun hydraulic cylinder piston to tilt the gun tube down and maintain the level position. If the tilt of the tank is more pronounced, the gyroscope precess deflection is increased and more resistor tabs touch each other, reducing the circuit resistance and increasing current flow to the control valves. The control valves adjust the pressure to the hydraulic cylinder that controls the gun angle, keeping it level. The second drawing shows the stabilizer system operating when the tank tilts down where the gyroscope precesses the opposite direction, engages a different set of resistors which pull the breech end of the gun down, maintaining the level position. The gyroscope spins at approximately 14,000 rpm.

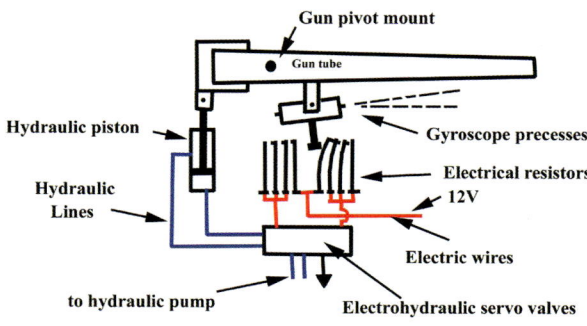

Stabilizer system operating when the front of the tank tilts up. (Roberts Collection)

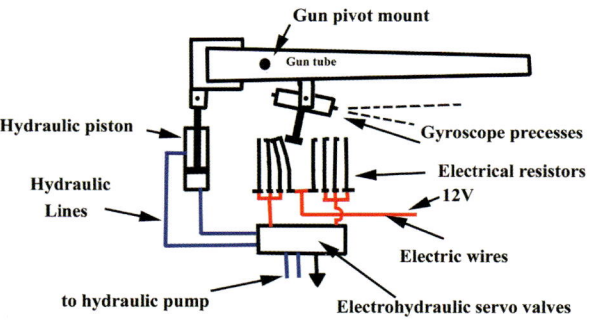

Stabilizer system operating when the front of the tank tilts down. (Roberts Collection)

(Top left) Inside the control box cover is the gyroscope and associated resistor network that operates the gun tube leveling piston. (Roberts Collection)

(Right) Field modified mono-gyro-control box with gyroscope and circuitry inside. (Roberts Collection)

(Bottom left) Control box that controls the leveling and gun firing. The lower toggle switch delivers power to the knob on the left, which adjusts the gun recoil. The knob on the right controls the stiffness of the leveling system. (Roberts Collection)

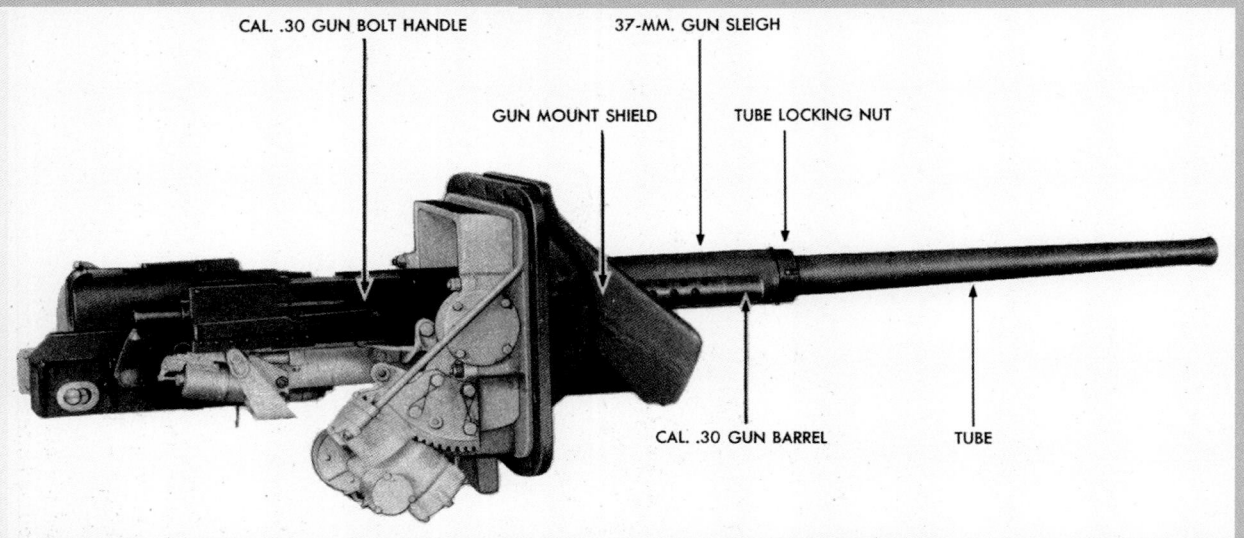

CAL. .30 GUN BOLT HANDLE

37-MM. GUN SLEIGH

GUN MOUNT SHIELD

TUBE LOCKING NUT

CAL. .30 GUN BARREL

TUBE

37-mm Gun, Tank M6 mounted in tanks. (FM 23-81)

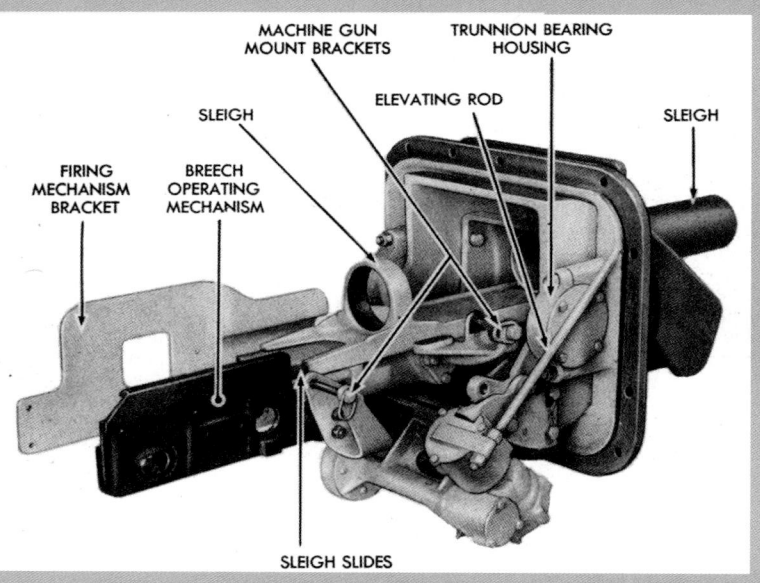

MACHINE GUN MOUNT BRACKETS

TRUNNION BEARING HOUSING

ELEVATING ROD

SLEIGH

SLEIGH

FIRING MECHANISM BRACKET

BREECH OPERATING MECHANISM

SLEIGH SLIDES

Right view of gun mount. (FM 23-81)

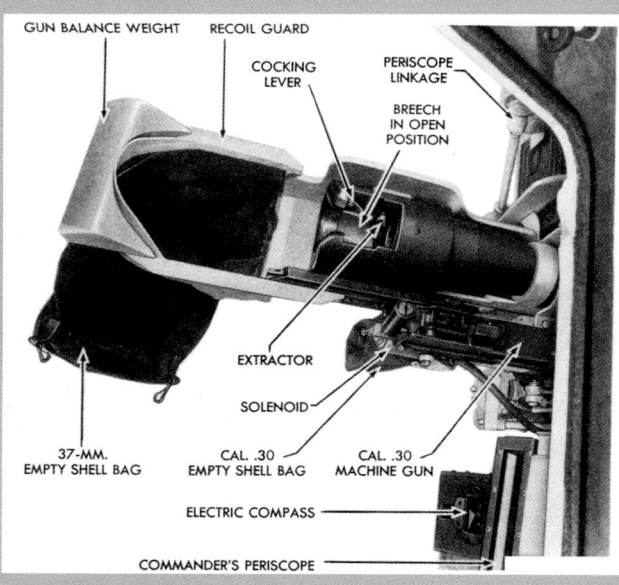

GUN BALANCE WEIGHT

RECOIL GUARD

COCKING LEVER

PERISCOPE LINKAGE

BREECH IN OPEN POSITION

EXTRACTOR

SOLENOID

37-MM. EMPTY SHELL BAG

CAL. .30 EMPTY SHELL BAG

CAL. .30 MACHINE GUN

ELECTRIC COMPASS

COMMANDER'S PERISCOPE

View of breech end of 37-mm gun, M6. (FM 23-81)

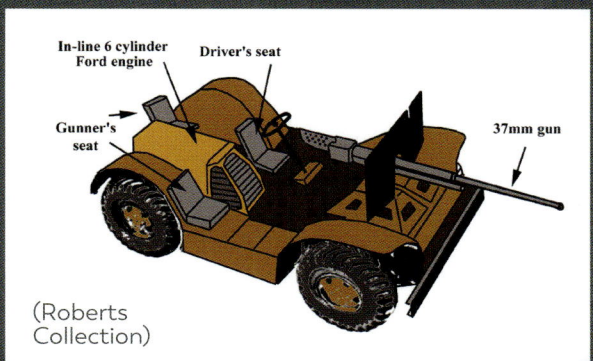

In-line 6 cylinder
Ford engine

Driver's seat

Gunner's
seat

37mm gun

(Roberts
Collection)

In Profile:
Ford T8
37-mm Gun
Motor Carriage

Known as the "Swamp Buggy," the T8 gun motor carriage was the first vehicle designed specifically to carry the 37-mm antitank gun. The vehicle was manufactured by Ford Motor Company using components from the standard 1½-ton truck (Burma Jeep). The standard Ford six-cylinder engine was in the middle of the vehicle to the right side of the driver. This allowed the gun to be mounted in the front of the vehicle. The gross vehicle weight was 4,520 lb with 100 rounds of ammunition and a three-man crew. Although this was intended to be an antitank vehicle, it offered very little protection for the crew. Development of this vehicle ceased in April 1942.

37-mm Automatic Cannon

There were three types of airborne automatic cannons, the M4, M9, and M10. These were designed for aircraft usage, were of the long recoil type and used belted ammunition. A solenoid fired the gun on command from a remote switch. The guns were primarily used in the Airacobra and Kingcobra aircraft, firing though the propeller spinner.

Specification: Automatic Cannons			
	M9	M4	M10
Caliber (bore) in	1,457	1,457	1,457
Firing rate, rpm	140	140	165
Muzzle velocity (HE) ft/sec	2,600	2,000	2,000
Muzzle velocity (AP) ft/sec	2,900	1,825	1,825
Weight of projectile (HE) lb	1.34	1.34	1.34
Weight of projectile (AP) lb	1.92 or 1.55	1.66	1.66
Weight of gun only, lb	405	213	231
Weight of gun tube only, lb	120	55	57.5
Overall length of gun, in	104	89.5	89.75
Interior ballistics number of grooves	12	12	12
Depth of grooves, in	0.02	0.02	0.02
Width of grooves, in	0.2314	0.2314	0.2314
Width of lands, in	0.15	0.15	0.15
Number of turns in the tube	1.56	1.56	1.56
Length of recoil, in	10.75	9.625	9.75

37-mm Automatic Gun M4

The 37-mm Gun M4 is a fully automatic, aircraft-mounted weapon using long recoil action with hydro-spring. The gun is mounted through the middle of the propeller spinner on P-39 and P-63 aircraft. It is fired using a solenoid and remote switch from the cockpit. Ammunition is fed through a 30-round endless belt. After a round is fired, the link that holds it in place remains on the link belt. After all 30 rounds are fired, the links must be manually reloaded. The gun was also adapted for usage on PT boats. The M9 and M10 guns were similar with slight variations such as spine locations, beefed-up longitudinal ribs, different solenoid locations, and characteristics as shown in the previous specifications.

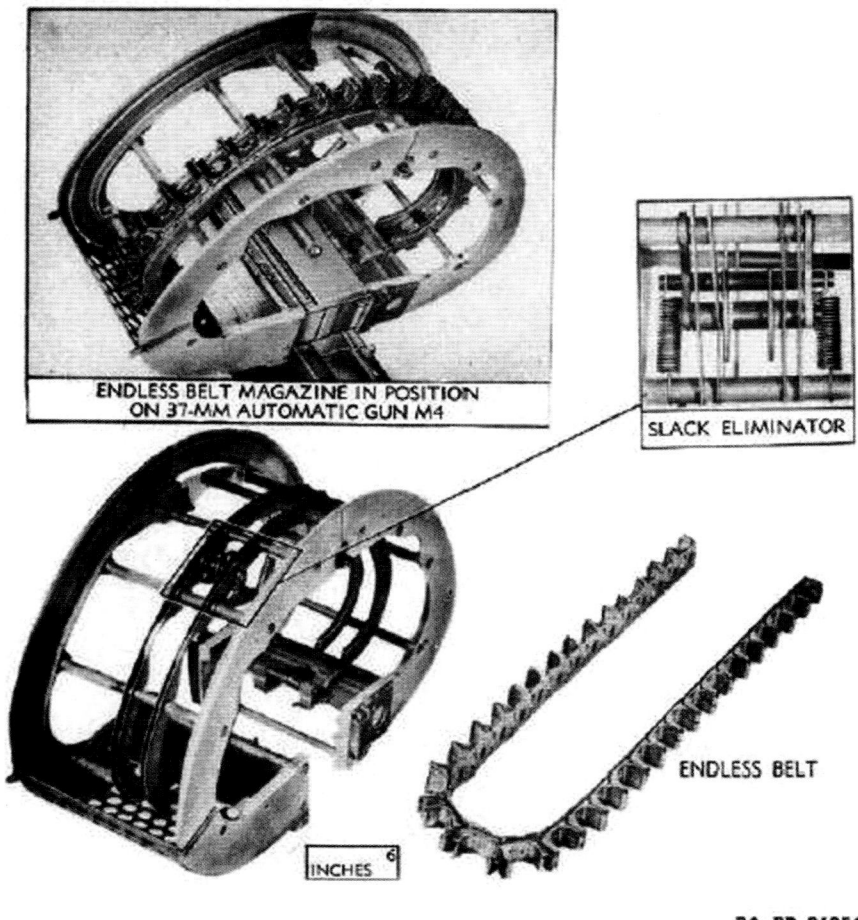

ENDLESS BELT MAGAZINE IN POSITION ON 37-MM AUTOMATIC GUN M4

SLACK ELIMINATOR

INCHES

ENDLESS BELT

RA PD 26854

Endless belt magazine M6. (TM 9-240)

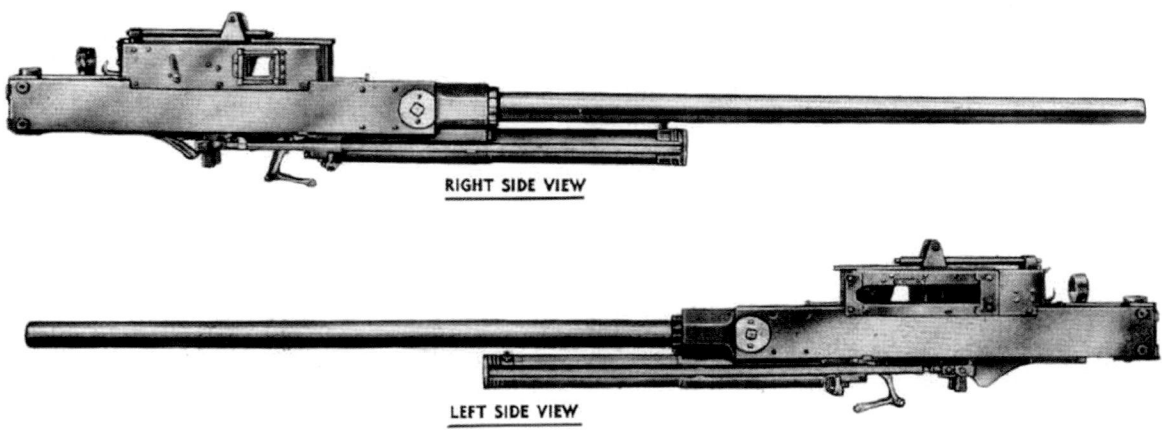

RIGHT SIDE VIEW

LEFT SIDE VIEW

37-mm Automatic Gun M4 used in aircraft and adapted to PT boats. (TM 9-240)

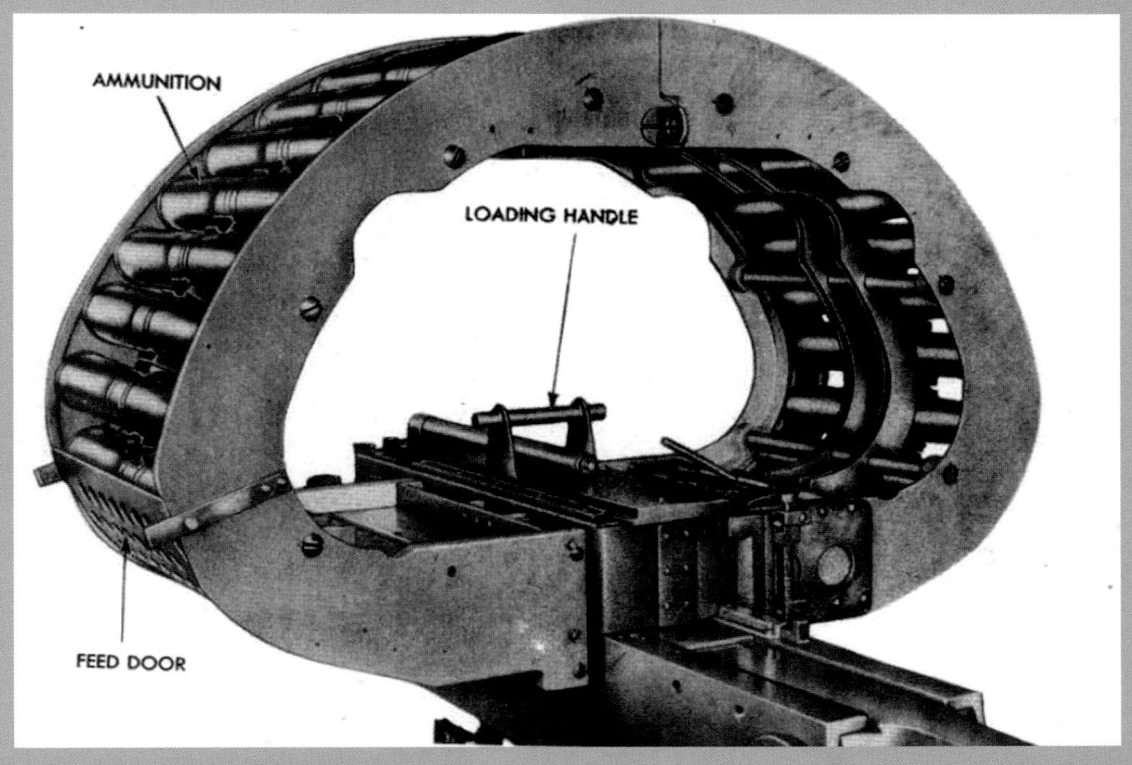

AMMUNITION

LOADING HANDLE

FEED DOOR

Ammunition belt for endless belt magazine mounted on gun. (TM 9-240)

37-mm Automatic Gun M9

The P-39 Airacobra was a fighter designed by Bell aircraft in the late 1930s and was one of the primary fighter aircraft in service when the U.S. entered WWII. The P-39 had an engine in the center of the aircraft with a long propeller shaft extending to the nose which contained a 37-mm gun. The aircraft was used by the Free French Air Force, the Royal Air Force, the Allied Italian Air Force, the Russian Air Force, and the U.S. in the South Pacific. The nose-mounted 37-mm gun made it easier for the pilots to aim at a target, since outboard mounted guns required the pilot to fire at a certain range to be effective.

In Profile:
P-39 Airacobra 37-mm Gun Location

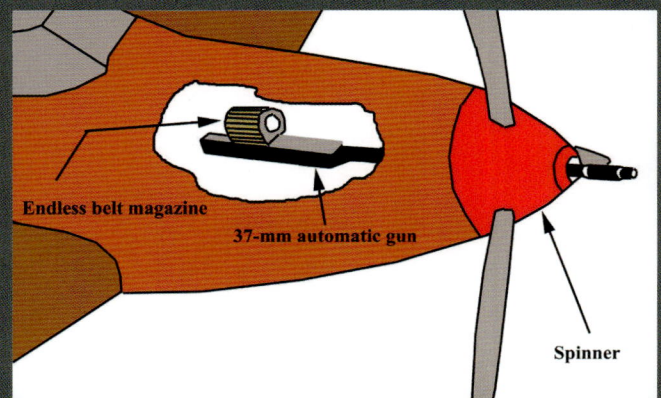

(Left) Cutaway of a P-39 Airacobra 37-mm gun location. (Roberts Collection)

(Below) 37-mm Automatic Gun AN-M10 mounted in an Airacobra. (Antina Richards-Pennock)

37-mm Automatic Cannon AN M10

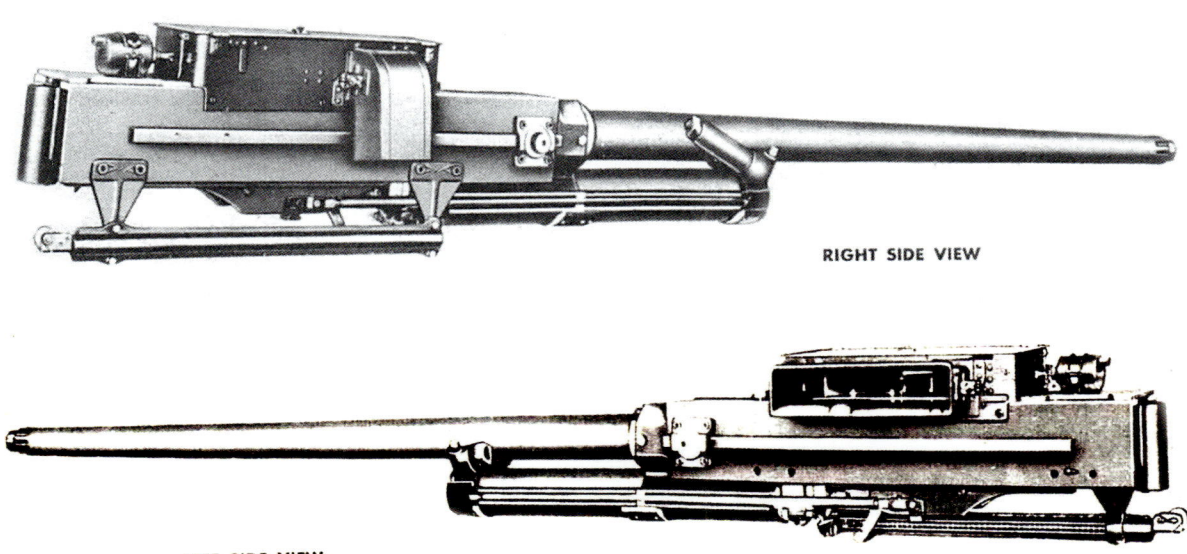

RIGHT SIDE VIEW

LEFT SIDE VIEW

M9 37-mm Automatic Gun. (TM 9-246)

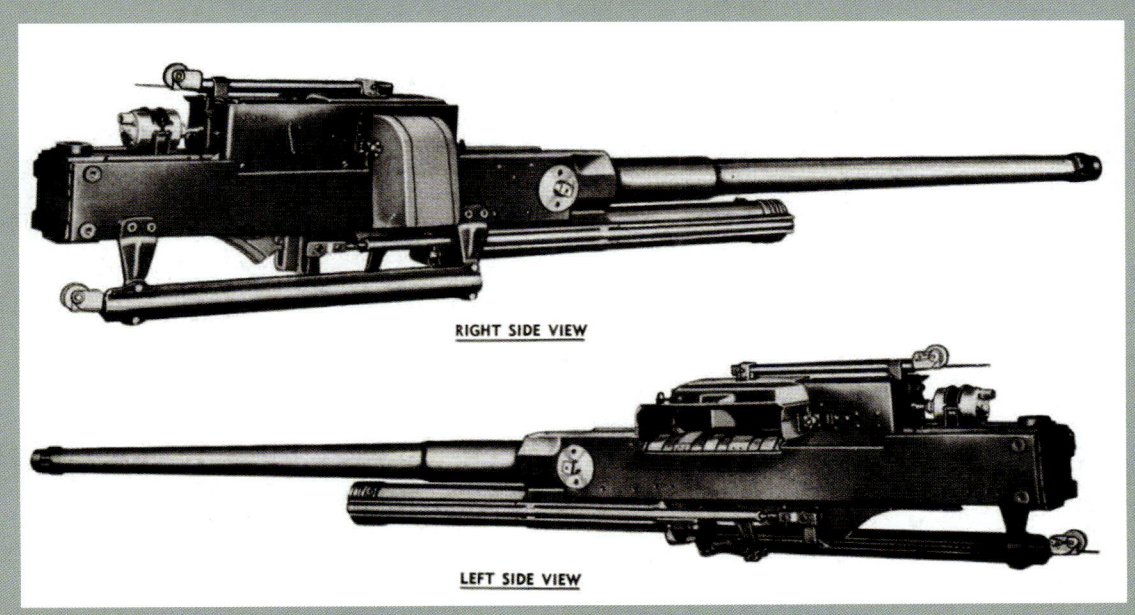

RIGHT SIDE VIEW

LEFT SIDE VIEW

37-mm Automatic Cannon AN-M10. (TM 9-240)

(Above) 37-mm T32 infantry gun. (U.S. Army)

(Facing page) Italy 1945, showing 37-mm T32 and 57-mm antitank guns on a carriage with the 10th Mountain Division. (U.S. Army)

37-mm Gun T32 and T33

Around late 1943, the Ordnance Department began development of a lightweight 37-mm gun that could be carried into the field by members of a rifle squad, designated the T32. This development was a result of Pacific theater experiences where a lightweight 37-mm gun would be helpful to forward units. The firing chamber was to accept the 5.69-inch M4 aircraft gun cartridge case with a working chamber pressure of 27,000 psi, firing an explosive projectile at 1,500 ft/sec.

A canister, anti-personnel cartridge was also expected to be used in the gun. The gun was to be mounted on a .50-cal machine-gun tripod that was modified to allow the traverse and elevation required for this gun. This was problematic in that the tripod was too light and the gun had to be repositioned after each firing because of the recoil. The total weight was expected to be approximately 250 lb. By July 1944, 155 guns had been sent to the Central Pacific with the understanding that if the gun was readily accepted by the troops, more would be on the way. The plan was to provide another 300 guns, but due to a lack of interest in using the gun among the troops in the field, the procurement was canceled. A number of guns were sent to the 10th Mountain Division in Italy in 1945.

In April 1944, the T33 gun started development as a light model with a weight not much different than the T32. By that time, the 57-mm M18 recoilless gun was ready. It provided a delivery of a much larger projectile and weighed approximately 44 lb. The recoilless gun M18 did not require a substantial breech since propellant gases escaped out the back, making this gun of higher caliber much lighter than the T32 or T33. The troops liked the 57-mm M18 recoilless gun which was used in Europe in March 1945 and in the South Pacific in June 1945. The T33 project was discontinued when the war ended.

T62 37-mm Recoilless Rifle

In July 1945, the Ordnance Department requested Frankford Arsenal Ordnance Laboratory to develop a 37-mm single-shot recoilless rifle, designated 37-mm T62. The recoilless gun was designed to fire the M54 and M63 projectiles at velocities of approximately 1,200 ft/sec. It was to be shoulder fired and used as an anti-personnel weapon. The prototype was developed and first fired in May 1946, well after the end of World War II. Modifications were made to the gun and the results were satisfactory, but the end of hostilities negated any further development, and the project was canceled.

| 3 37-mm Ammunition Usage in World War II

The U.S. 37-mm gun used a variety of ammunition types, such as armor-piercing, high explosive, and canister shot, for different applications. The common components on the 37-mm cartridge are shown in the drawing below. The primer contains a volatile chemical that ignites when struck by a firing pin, sending a flame into the cartridge case containing the propellant. This ignites the propellant, causing high-pressure rapid burning that sends the projectile out the barrel on the way to the target. Upon contact or after a period of time, the fuze ignites, initiating the explosive charge, destroying the target.

A projectile is a missile propelled by the force of expanding gases. The projectile can be either solid which destroys a target by transfer of kinetic energy (energy of motion) or contains an explosive chemical that destroys the target by the blast. Early projectiles were round balls that behaved erratically when fired. By about 1860, the rifled cannon was developed and projectiles became cylindrical with a pointed nose.

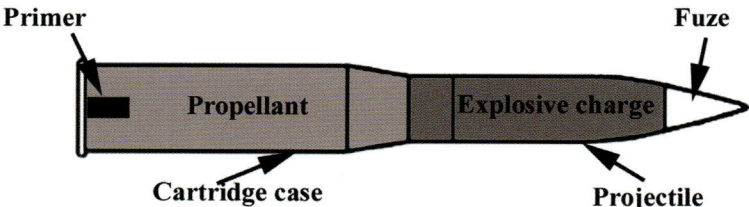

Typical 37-mm cartridge design of World War II. (Roberts Collection)

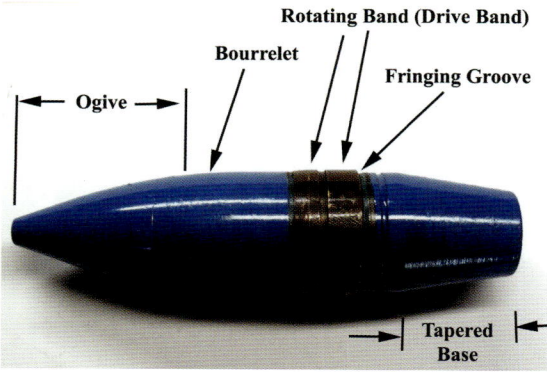

Typical parts of a 37-mm projectile. (Roberts Collection)

A typical 37-mm projectile is shown left. The frontal nose is called the ogive, from the French meaning nose cone. The ogive curve has a radius that describes the arc of the ogive, usually expressed in terms of calibers. The caliber is the diameter of the gun tube between lands at opposite ends. The rifling in a gun tube has grooves and lands, the lands being the raised portion of the gun tube and the grooves being the spaces between the lands. Just behind the ogive is the bourrelet, which is French for sealing strip.

The bourrelet helps stabilize the projectile as it exits the gun tube. Its purpose is to reduce the tendency of the projectile to pitch and yaw. It is precisely machined to fit in the gun tube without forming any grooves as the projectile exits the gun. The longitudinal width of the bourrelet is typically 1/6 caliber and machined with a smooth surface to reduce gun tube wear. In the machining of the bourrelet, some clearance is allowed to account for variations in the diameter of the gun tube. The rotating band or drive band is a soft metal ring swaged into the projectile. Swaging is a form of pressing soft metals into a harder core. When the gun is fired, the lands in the gun tube engage in the soft metal and engrave slots that match the lands, thereby causing rotation of the projectile, which adds stability to the flight and increases accuracy. The rotating band fills the grooves in the gun tube, sealing off any avenue for escaping propellant gas. The fringing groove is just behind the drive band and is a collecting area for metal debris that is a result of the engraving. This reduces the buildup of debris in the grooves, which can affect accuracy. Copper and gilding metal are typically used for the drive band. Gilding metal is a form of brass with much higher copper than zinc content. The rear of the projectile is tapered to make the shape more aerodynamic with less drag. This is called the boat-tail design and is typically a 6° to 8° taper. A base plate may be installed to cover the empty cavity of an explosive projectile. Projectiles are typically coated with paint to prevent corrosion. There are certain color codes used: yellow is high explosive, red is low explosive and shrapnel, solid kinetic rounds are painted black, practice ammunition is painted blue, and chemical rounds have a blue-gray color with circumferential bands painted on the round representing the type of chemical, such as white phosphorus smoke.

Projectiles used during World War II for the 37-mm gun were constructed of forged steel with either a solid body or hollow filled with an explosive. The explosive projectiles were equipped with a fuze that would burst on impact or burst shortly after impact (time delay).

There were three major categories of 37-mm projectiles used in World War II: armor-piercing, high explosive, and canister. Armor-piercing projectiles have forged steel bodies which are solid or hollow to accept an explosive. A steel armor piercing cap is typically placed at the front of the projectile. This type of projectile is referred to as APC (armor-piercing, capped) projectile. The cap tended to attenuate the force of impact so that the projectile body would not fracture prior to penetrating the armor. Ballistic caps were developed to add a more aerodynamic ogive, reducing drag and increasing range. Finally, the softer metal cap tended to easily grab the target metal when fired at an angle, reducing the chance of a ricochet.

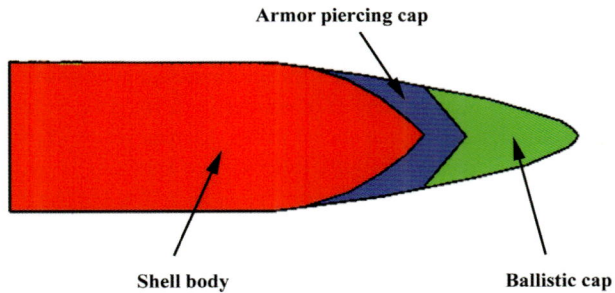

APCBC projectile with armor-piercing cap and ballistic cap. (Roberts Collection)

The point of the cap is made of a high-carbon chrome steel and locally heat treated such that the point is very hard while the rest of the cap is softer and resistant to shattering. The cap places very high stresses in the face-hardened armor, preparing the way for the projectile body which actually does the penetrating of the armor plate.

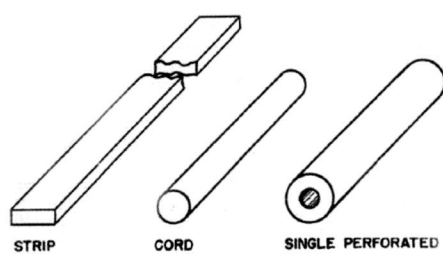

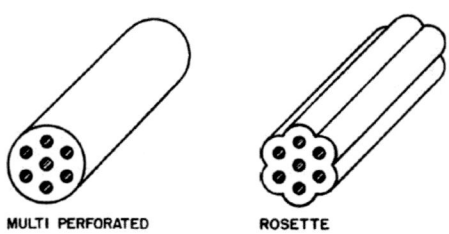

Gunpowder grain shapes. (TM 9-1904)

The 37-mm gun was also equipped with a canister round containing steel balls that would scatter as soon as they left the muzzle. This is a typical anti-personnel round similar to a shotgun shell where pellets scatter and strike objects over a wide area.

Fuzes are devices that set off the explosive material in a projectile. There were two main types of fuzes used in 37-mm ammunition in World War II: point fuzes at the front of a projectile and base fuzes at the base of a projectile.

During World War II, there were two classes of military explosives: low-order explosives and high-order explosives. Low-order explosives are characterized by flame or spark initiation, deflagration reaction and resulting displacement of a target. Deflagration is a form of rapid combustion. High-order explosives are characterized by mechanical shock initiation, detonation reaction and shattering of the target. A high-order detonation results in very high-velocity pressure waves in excess of 20,000 ft/sec. A low-order detonation results in much lower pressure waves, on the order of the speed of sound. Low-order explosive powders are typically used for propellants. The high-order explosive substances are typically used for bursting charges in a projectile and not as propellant because their pressure rise is so rapid that gun damage could occur before the projectile ever leaves the gun.

Nearly all the ammunition used in the 37-mm gun used FNH powder as the propellant. FNH stands for flashless non-hydroscopic. This means that the propellant has a significantly reduced flash when the projectile is fired, therefore reducing the chance of alerting the enemy as to the gun's position, and the propellant does not absorb moisture, which would decrease performance. This is often referred to as smokeless powder.

FNH powders are mixtures of nitrocellulose and material additives that reduce the flash and the propensity to absorb moisture. FNH powder was used in most guns of caliber 37 mm and higher in World War II. The flashless feature of this propellant varied between different guns which helped reduce the flash, but did not totally eliminate it. Nitroglycerin was added to certain FNH formulations for use in smaller guns where rapid burning was required along with high-velocity requirements. FNH is supplied in powder form, but the powder grain shape is not spherical in nature. They may have complex shapes like that shown in the diagram above. These shapes tend to control the burning rate of the propellant. When the grains burn, their surface area may be reduced and the burning rate decreases. This happens with the strip, cord, and other nonperforated grains. For the perforated grains, the surface area during burning does not substantially decrease, resulting in an increased burning rate. These grain shapes are adjusted for a particular gun so that the pressure rise during firing does not exceed the capacity of the gun.

The development of FNH powder was a result of the desire to not use black powder as a propellant. Black powder had been used for years as a propellant. It is a mixture of saltpeter, sulfur and charcoal. The problems with black powder were: the flash and smoke from the gun was great, revealing the gun crew's position and subsequent discovery by the enemy; black powder caused excessive wear in the gun barrel; hot chamber residue after a shot made it dangerous to load a subsequent round; the rate of burning was difficult to control; black powder absorbs moisture readily, making the burning rate vary significantly; and finally, black powder is easily initiated by flame, spark or friction, making it dangerous to handle. This led to the development of nitrocellulose powders (NC). Early on, two major problems were noted with NC powders: firstly, they easily absorbed water and deteriorated by hydrolysis. The second was a bright muzzle flash that occurred during firing. Materials to control these problems were added to NC powder, resulting in the FNH powder. Diphenylamine (DPA) adds stability to the powder by combining with nitrous

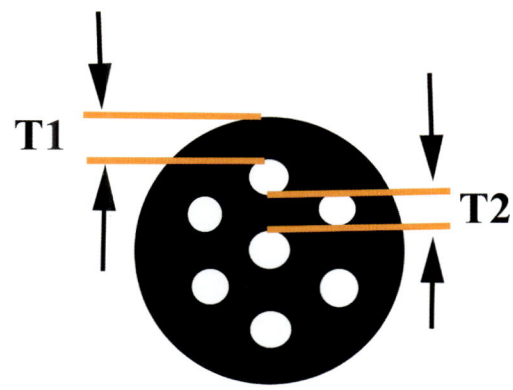

Multi-perforated grain structure. (Roberts Collection)

fumes from deterioration. Acidic gaseous oxides of nitrogen develop as a result of deterioration, releasing heat (self-heating) that results in a temperature rise that can reach auto-ignition temperatures. The DPA combines with these acidic fumes, reducing the rate of deterioration, prolonging the life of the propellant powder and enhancing its stability. Another ingredient added to FNH powder is dibutyl phthalate (DBT) which is a cooling agent and inhibits the absorption of moisture. This substance does not explode, is essentially inert, and cools the propellant gases below their kindling point.

Burning rate of the FNH powder is controlled by the shape of multi-perforated grains. The averages of the distances T1 and T2, as shown in the diagram Rosette grain structure (upper right), control the rate of burning. The thinner the average T, the quicker the propellant burns. The thicker the average T, the slower it burns. Typically, the larger-caliber guns have grains with the thicker value of T. This allows time for the projectile to exit the long, high-caliber gun tube. One exception is for short-barrel, high-caliber guns such as the M8 howitzer. In those cases, quicker burning is desired since the time the projectile is in the barrel is short. For the 37-mm ammunition, the FNH powder grain configuration is adjusted to yield a certain velocity consistent with the maximum chamber pressure allowed in the gun chamber.

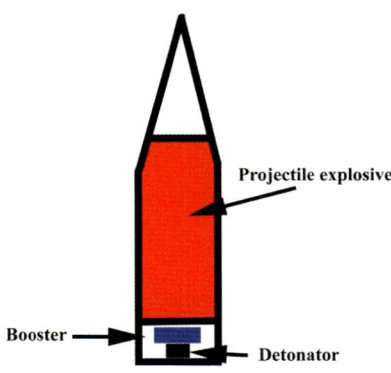

Booster charge. (Roberts Collection)

The primer/detonator is a device at the end of the cartridge case that sets off the propellant when struck by a firing pin. Mercury fulminate had been used as a primer explosive but lead azide, tetryl, and other chlorates have replaced mercury fulminate. Mercury fulminate is very sensitive, highly explosive and detonates easily with friction, spark flame or shock. Tetryl, which is short for 2,4,6-trinitrophenylmethylnitramine is an explosive commonly used in projectiles in World War II as the main explosive and also as a booster. A booster is an auxiliary, more sensitive explosive that amplifies the initiation of the detonator to assure that the main projectile explosive explodes. Many high-energy explosives such as TNT (2,4,6-trinitrotoluene) have low sensitivity to initiation but explode with high energy with a booster charge. Tetryl was often used as a booster charge as well as the main explosive charge in projectiles during World War II.

U.S. 37-mm Ammunition Used in World War II

The typical American 37-mm cartridge has three major parts: the projectile, cartridge case, and primer. The projectile is the business end of the cartridge since it strikes the target kinetically or chemically (explodes). A kinetic round relies on momentum to damage the target. The high-velocity energy or kinetic energy is imparted to the target, causing damage or destruction. The explosive round relies on a violent chemical reaction (the explosion) to defeat a target. The cartridge case contains the propellant (FNH) that drives the projectile out the gun tube on the way to the target. The primer is a small container of a volatile mixture that when struck with a firing pin, ignites and sets off the propellant charge. Brass is typically used for cartridge cases since it is relatively soft, easy to form, seals the firing chamber well and extracts easily from the firing chamber. Brass is also corrosion resistant. For the higher-velocity rounds, the cartridge case is tapered to 37 mm, allowing a large volume for the propellant.

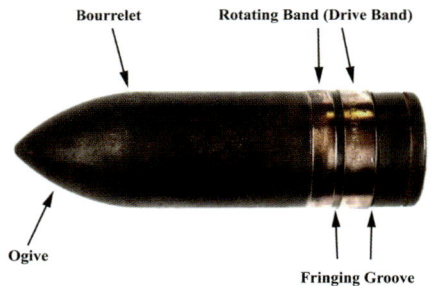

37-mm M74 projectile. It is approximately a 2-caliber ogive, a blunt design. Ogive is a pointed curved surface that forms the nose cone of a projectile. The bourrelet is an accurately machined area near the nose that just fits the lands on the gun. It tends to reduce the pitch and yaw motion of the projectile when leaving the gun tube. (Roberts Collection)

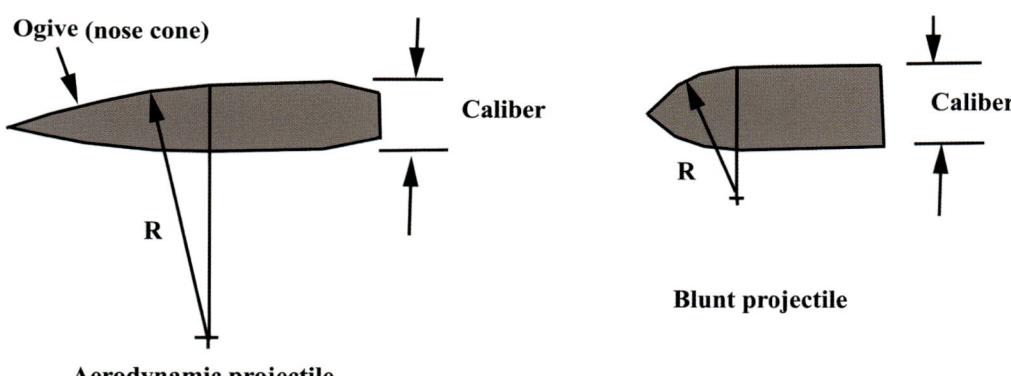

Projectiles came in many shapes and sizes for the American 37-mm guns. Ogive, which is French for nose cone, is the shape of the front of a projectile. The caliber of an ogive is the radius (R) of the shape of the ogive curve divided by the diameter (caliber) of the projectile, i.e., R/caliber. The projectile at the left above, which also shows a tapered base, is approximately a 5-caliber ogive. The blunt projectile at the upper right is approximately a 2-caliber ogive. It is shown with a flat base. The higher the caliber ogive number the more streamlined the projectile and higher-impact velocity due to less aerodynamic drag. (Roberts Collection)

(Top left) Square base 37-mm Mk II projectile with an approximate 2-caliber nose ogive. This is considered a blunt nose projectile. The drive band has been engraved, indicating that this projectile was fired through a gun. The fuze is in the base, i.e., base detonating. (Roberts Collection)

(Top right) End of a cartridge case showing hole for primer. (Roberts Collection)

(Middle) Inside view of primer. (Roberts Collection)

(Bottom left) Side view of primer M23A2. The primer contains a more volatile compound that is used to set off the propellant in the shell. The dent in the middle of the end is from the firing pin, indicating that this primer was in a cartridge that had been fired. (Roberts Collection)

(Top) 37-mm M17 brass cartridge case. (Roberts Collection)

(Bottom) 37-mm cartridge, disassembled, manufactured in 1942. (Roberts Collection)

In Profile:
Field-manufactured Model Training Gun & M3 Antitank Gun

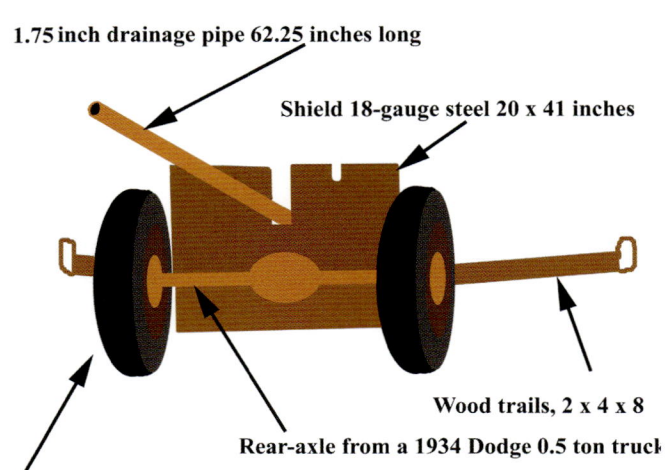

1.75 inch drainage pipe 62.25 inches long

Shield 18-gauge steel 20 x 41 inches

Wood trails, 2 x 4 x 8

Rear-axle from a 1934 Dodge 0.5 ton truck

Tires from a 1934 Dodge 0.5 ton truck

(Left) A field-manufactured model training gun, based on the M3 antitank gun. There was such a shortage of weapons prior to World War II that practice guns were used very much like broom sticks for infantry practice rifles. These guns varied in construction depending on the Army unit making them and what materials were available. In this design shown on the left, a salvaged rear axle assembly and tires from a truck acted as the wheel assembly. A 62.25-inch length of 1.75-inch steel drainage pipe approximated the gun tube. A practice cartridge could be inserted into a wooden breech at the end of the pipe. The trails were 8-foot 2x4s and could be spread out into the firing position using the lunettes at the ends. (Roberts Collection)

The training gun on the left was to act as a practice aid for crewing a gun like the M3 shown below. (Roberts Collection)

37-mm Gun M1916

This gun was introduced in World War I, but the projectile was low in velocity and could not penetrate more than 0.375 inches of armor at 500 yards. In World War II, the M1916 was used primarily as a sub-caliber gun for aiding in the aiming of large-caliber weapons such as the 75-mm and 155-mm howitzers and guns for training purposes. The primary ammunition used in the M1916 was the Mark I cartridge.

Cartridge, Fixed, Mark I

The complete round is designated Shell, Fixed, HE Mark I. Fixed indicates that the cartridge is provided with all parts, propellant, case, projectile, primer, and fuze in one unit. It is assembled as one unit that is inserted into the breech of a gun. (Non-fixed ammunition, which is shipped in units with the projectile and propellant in separate pieces, is used in large-caliber guns where the projectile is rammed into the firing chamber and the propellant is loaded separately.) The propellant amount used in this cartridge during World War II was 550 grains (1 lb = 7,000 grains) of FNH powder, yielding a muzzle velocity of 1,312 ft/sec, a maximum range of 4,500 yards, and an effective range of 1,200 yards. The projectile is about 3.56 inches long with a 2.25-caliber ogive. The projectile weighs approximately 1 lb, hence the nickname of the 37-mm gun was the "one pounder." It is filled with 0.034 lb of black powder. The cartridge is approximately 6.371 inches in length and weighs approximately 1.49 lb. A sand-filled Mk I was manufactured for training purposes. There was also a canister-shot Mk I, which was essentially a shotgun shell containing 38 individual lead balls. When the gun was fired, the shock ruptured the canister case and the lead balls shot out of the muzzle at about 1,200 ft/sec with a maximum effective range against personnel of 75 yards.

Complete round, 37-mm high explosive for 37-mm gun M1916 manufactured 1942. (Roberts Collection)

37-mm Gun M3

Cartridge, Fixed, Mark II

The Mark II cartridge, used in the 37-mm gun M3, is a substitute for armor-piercing shot against lightly armored targets since the maximum armor plate penetration of the round is 0.375 inches at 500 yards. The complete round Mk II HE cartridge was considered obsolete and superseded by the M63 in 1942. The M16 cartridge case and the M23A1 primer were standard issue for these high-explosive projectiles. Total weight with fuze was approximately 1 lb. The fuze is base-detonating M58 and is standard for this ammunition. The propellant charge is 550 grains of FNH powder loosely packed into the cartridge. The powder is poured into the cartridge case rather than compacted into the cartridge case. The projectile is made of bar steel and has an explosive charge of 0.06 lb of TNT. It is 4.45 inches long with a 2.25-caliber ogive. The detonator assembly contains a priming mixture of lead azide and tetryl. The cartridge is 6.92 inches long and weighs approximately 1.61 lb.

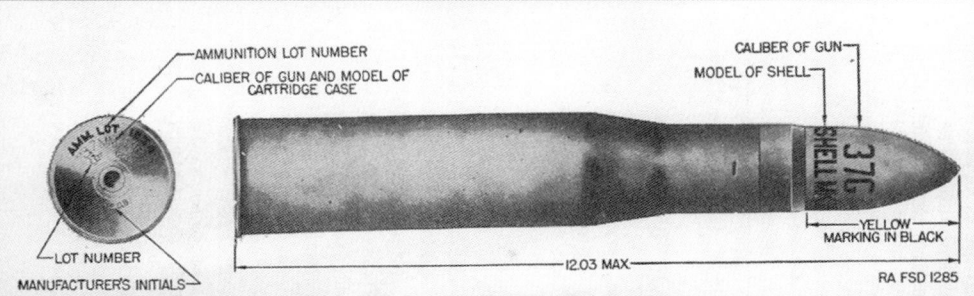

High-explosive Mk II cartridge with fuze base-detonating M38A1. (C. Roberts, *U.S. Airborne Tanks 1939–1945*)

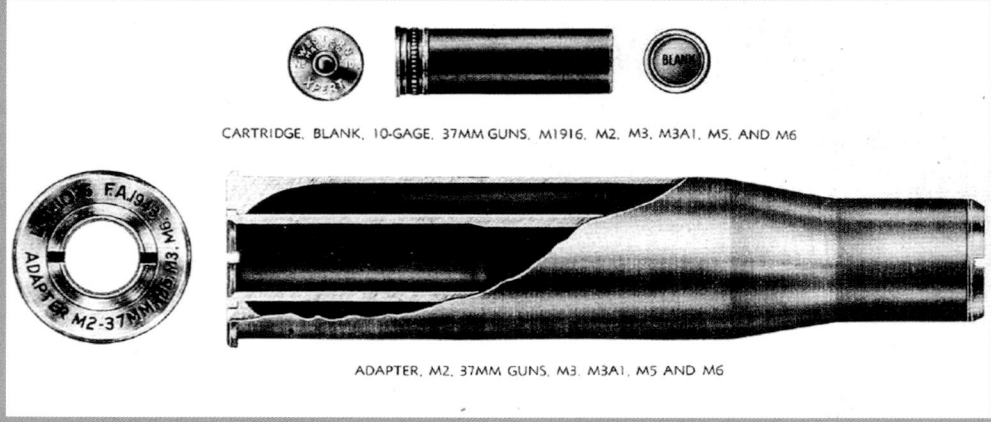

Adapter cartridge M2 for training with a shotgun shell for simulated fire. The shotgun shell is a blank and is inserted into the end of the adapter and is used for training, thus conserving valuable ammunition. (C. Roberts, *U.S. Airborne Tanks 1939–1945*)

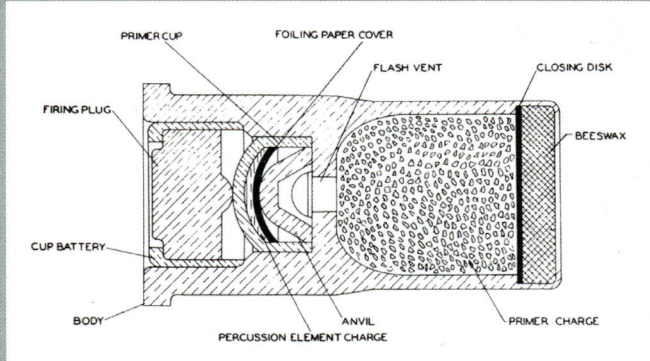

Primer, percussion, M23A2. The firing pin shocks the primer charge, resulting in ignition. (TM 9-1904)

Cartridge, Fixed HE M63

The M63 HE cartridge was the standard explosive round used on the M3 antitank guns and the M5 and M6 tank guns. The M23A2, 20-grain, percussion primer was standard on these rounds. The propelling charge is approximately 0.44 lb of FNH powder packed loosely (not compacted) into the cartridge case, yielding a muzzle velocity of 2,650 ft/sec. The high-explosive shell M63 weighs 1.61 lb and is 5.92 inches in length. The projectile ogive radius is 13 inches or about a 9-caliber ogive. The explosive charge is 0.085 pounds of flaked TNT, packed so as to surround the booster cavity (see below). The fuze is base-detonating M58. The complete round is 14.09 inches in length and weighs approximately 3.03 pounds.

M63 MOD.1 with brass cartridge case MK1A2 for 37-mm sub-caliber gun only. (Roberts Collection)

Shell HE M63 for 37-mm Guns M3, M3A1, M5, and M6. (TM 9-1904)

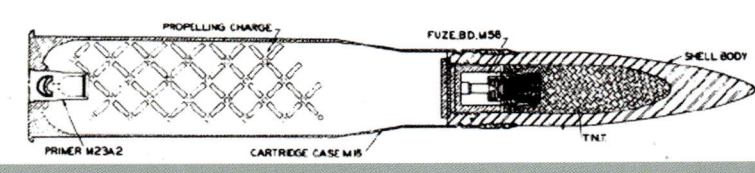

High-explosive M63 cartridge with steel cartridge case M16B1. Steel case was used as a substitute for brass, but there were problems. Reducing corrosion was a problem with the steel case. The steel case had to be coated with a varnish to reduce corrosion. Brass did not require such a coating. The brass is much more compliant than steel, making the brass cartridge easier to extract. When the steel expands during firing, it may not return to its original shape as easily as brass, making it difficult to extract. Consequently, brass casings were desirable. (Roberts Collection)

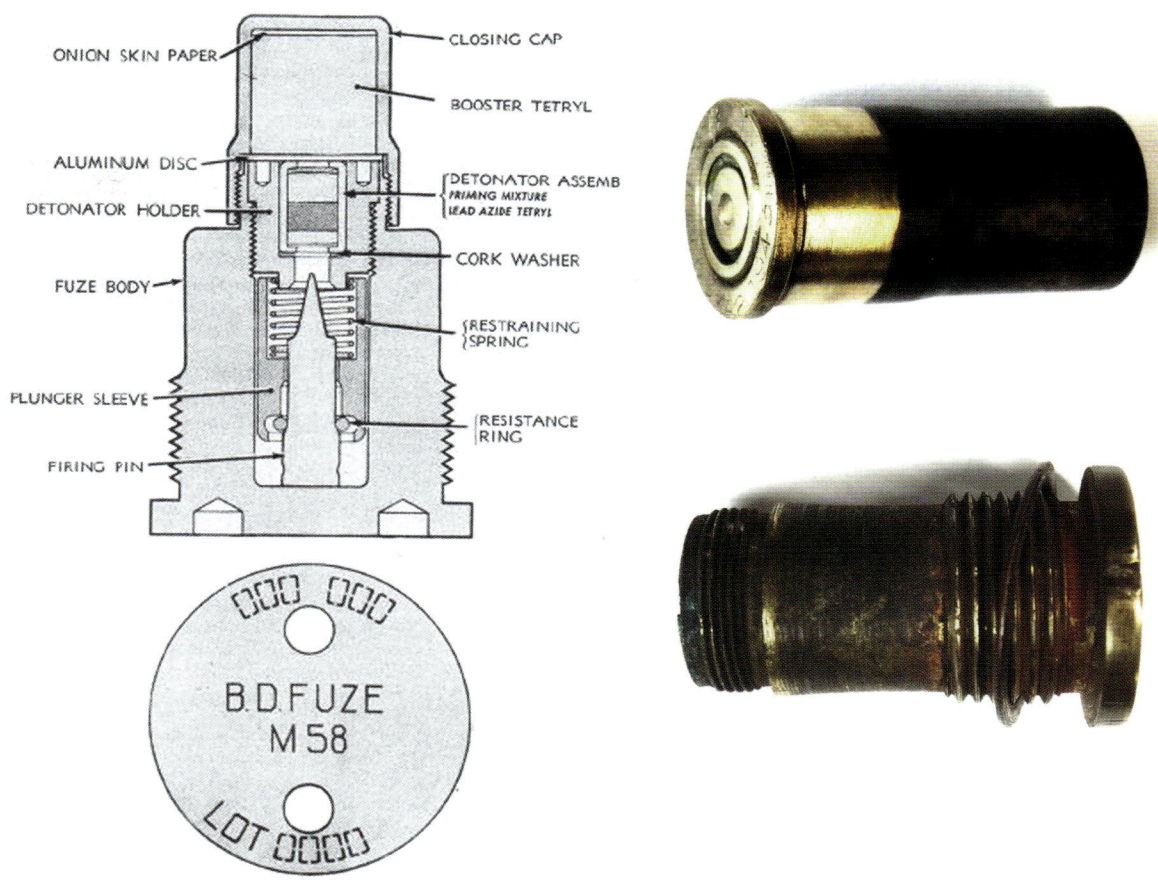

ONION SKIN PAPER
CLOSING CAP
BOOSTER TETRYL
ALUMINUM DISC
DETONATOR ASSEMB
PRIMING MIXTURE
LEAD AZIDE TETRYL
DETONATOR HOLDER
CORK WASHER
FUZE BODY
RESTRAINING
SPRING
PLUNGER SLEEVE
RESISTANCE
RING
FIRING PIN

OOO OOO
B.D. FUZE
M 58
LOT OOOO

(Left) Base-detonating fuze M58 for HE shell M63. When the cartridge is fired, the acceleration causes the plunger sleeve to move downward and engage the groove in the firing pin This exposes the firing pin so that upon impact, it moves forward initiating the priming mixture and setting off the charge. (TM 9-1904)

(Top right) M23A2 primer which has been fired as evidenced by the center punch dent by the firing pin at the middle of the base. The denting action by the firing pin causes a sudden shock that sets off the primer. (Roberts Collection)

(Bottom right) Base-detonating fuze M58 for HE shell M63. The booster cap at the top has been consumed, indicating that this fuze has fired. The priming charge is a mixture of tetryl and lead azide. The booster charge is tetryl. (Roberts Collection)

Cartridge Fixed, APC (Armor-piercing, Capped), M51 With Tracer

This cartridge is an armor-piercing shot, effective against all armor plate materials. The cartridge case M16 is standard on this round, with the M16B1 as a substitute steel case cartridge. The primer is the M32A2, 20-grain, percussion primer which is standard for most 37-mm rounds. The propellant is FNH powder (8.1 ounces) loosely packed, resulting in a muzzle velocity of 2,900 ft/sec. The projectile is hard steel with a soft steel cap that reduces the possibility of a ricochet and aids the hard steel penetrator in engaging the armor plate, especially at an angle. There is a cavity at the base that contains the tracer material which burns for approximately 2,000 yards of the flight of the projectile. The projectile weighs 1.92 lb and is 6.36 inches long with the aerodynamic cap attached. The complete round is 14.47 inches long and weighs 3.41 lb.

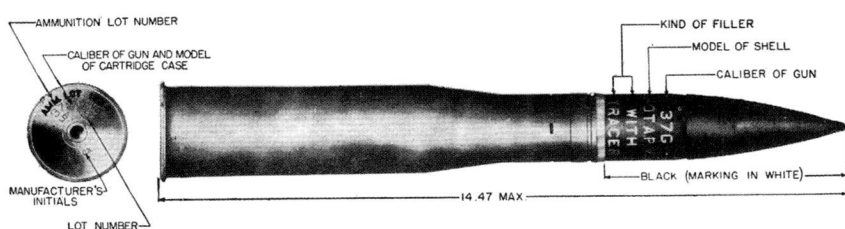

Cartridge, armor-piercing, M51 with tracer. (FM 23-81)

Target-practice M51 Cartridge With Tracer

This cartridge is designed for target practice as well as general field practice. The primer is the standard M23A2, 20-grain, percussion primer. The cartridge case is filled with 6.6 ounces of FNH powder loosely packed. The projectile is the same as the APC M51 and is a solid piece of steel that is not heat treated. No explosive is in the projectile.

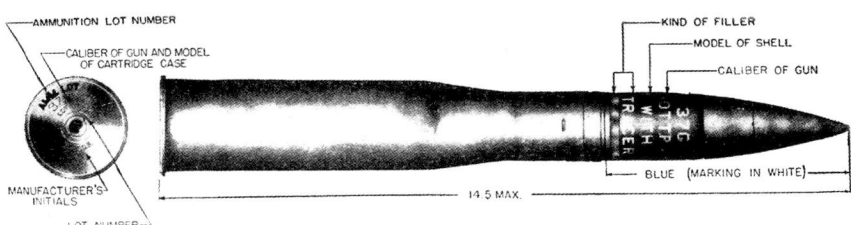

Target-practice M51 cartridge with tracer. (C. Roberts, *U.S. Airborne Tanks 1939–1945*)

Drill Cartridge M13

This drill cartridge is made of inert components and used components from service rounds. It is designed for training personnel in the handling of ammunition and in loading and unloading weapons. It has a hole in the cartridge case to verify that no propellant is supplied with the round. The loader would insert the round into the breech and close the breechblock, and then open the breechblock to retrieve the round.

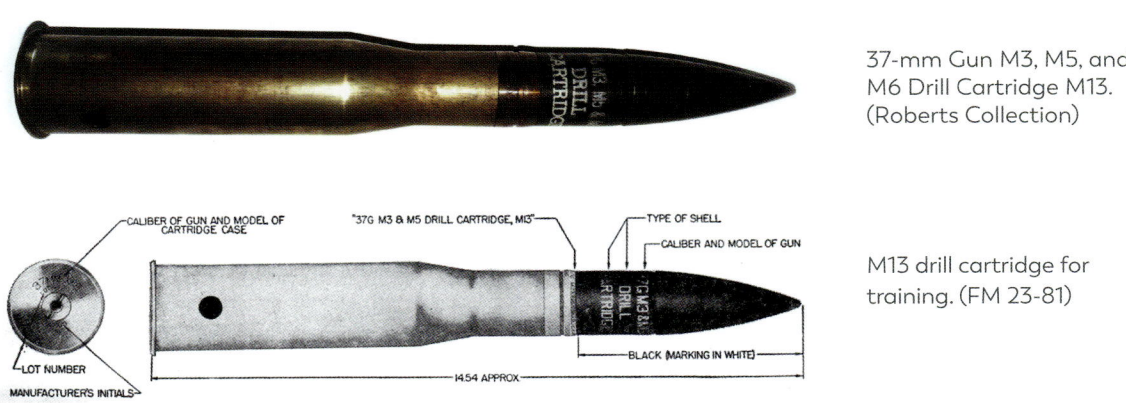

37-mm Gun M3, M5, and M6 Drill Cartridge M13. (Roberts Collection)

M13 drill cartridge for training. (FM 23-81)

37-mm Automatic Gun M1A2 (Antiaircraft)

The M1A1 and M1A2 are fully automatic weapons and require a 37-mm casing with an extraction groove at the base. There are three types of ammunition for the antiaircraft gun: high explosive (tetryl filler), armor-piercing, and practice. The cartridge case is the M17, which is nearly identical to the M16 cartridge case with the exception that there is an extraction groove at the base of the M17, while the M16 has an extracting flange. The M17B1 is identical in shape to the M17 with the exception that the M17B1 is made of steel. The M23A2, 20-grain, percussion primer is the standard for the antiaircraft ammunition.

Cartridge, Fixed HE, M54 With Tracer

This projectile was used on the M1A2 antiaircraft gun and the M4 aircraft gun (Airacobra). The cartridge case for the M1A2 antiaircraft gun was standardized as the M17 brass case or the M17B1 steel case, utilizing the M23A2 primer. The propellant was 6 ounces of FNH powder yielding a muzzle velocity of 2,000 ft/sec. The projectile M54 weighs 1.34 lb and is approximately 5.9 inches long. The length of the cartridge is 12.81 inches. The fuze is point-detonating M56. The projectile is machined from bar steel and is 4.13 inches long. A bursting charge of 0.1 lb of tetryl is pressed into the body in two pieces, a base pellet and main charge.

After the cartridge is fired, the self-destroying tracer eventually burns through to the main charge, thereby setting off the explosive. The projectile destroys itself after reaching 3,500 yards vertically or 4,000 yards horizontally. The actual limit of travel is much higher but the self-destroying aspect of the projectile limits the range. This is a safety aspect of the round so that if a projectile misses a target, it destroys itself before hitting the ground, thereby reducing the possibility of unintended casualties. The fuze is sensitive and will detonate when striking light materials such as aircraft structures.

The M4 gun used the MKIIIA2 cartridge case and the same M23A2 primer. The length of the cartridge is 9.75 inches. The propellant is 2.4 ounces of FNH powder yielding a muzzle velocity of 2000 ft/sec.

Shown right is the M56 point-detonating fuze. When the cartridge is fired, it begins to spin which, by centrifugal force, causes the eccentric slider to move against the slider spring. This then aligns the slider charge with the firing pin and detonator train, thereby activating the projectile. This is a safety device since the point detonator is very sensitive and if the ammunition is accidently dropped, it could detonate if no safety device were present. The reason for the sensitivity of the detonator is to be able to be set off when striking the relatively light materials found in aircraft. When the point detonator strikes the aircraft, the firing pin ignites the first explosive in the detonator train, setting off the booster charge.

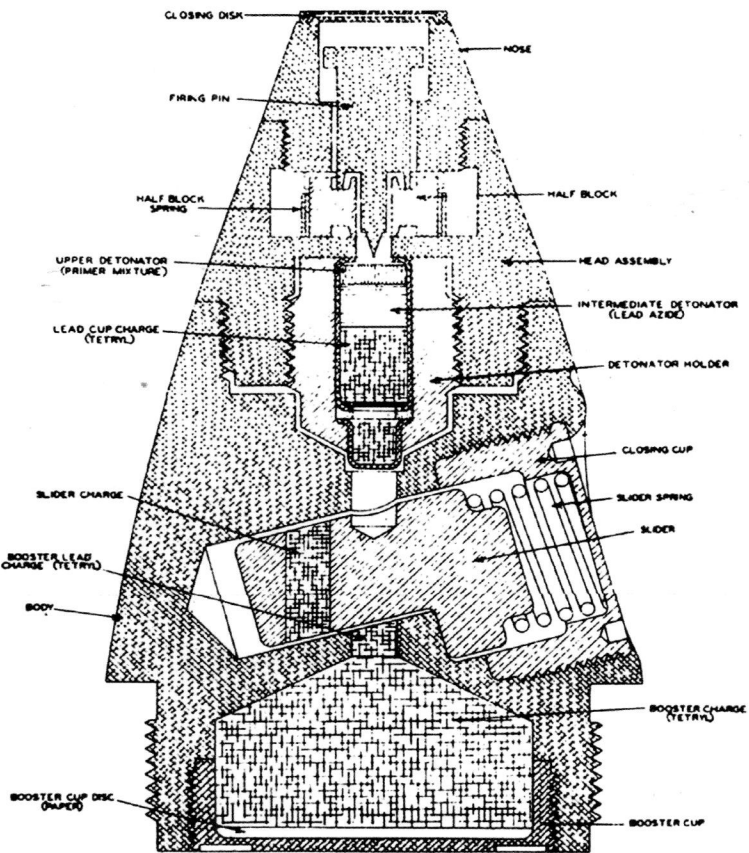

Fuze, point-detonating M56. (TM 9-1904)

(Left) Cartridge HE M54 for 37-mm guns M1A2, M4 and M9. (TM 9-1904)

(Below) High-explosive cartridge M54 for the aircraft guns M4 and M10. (TM 9-240)

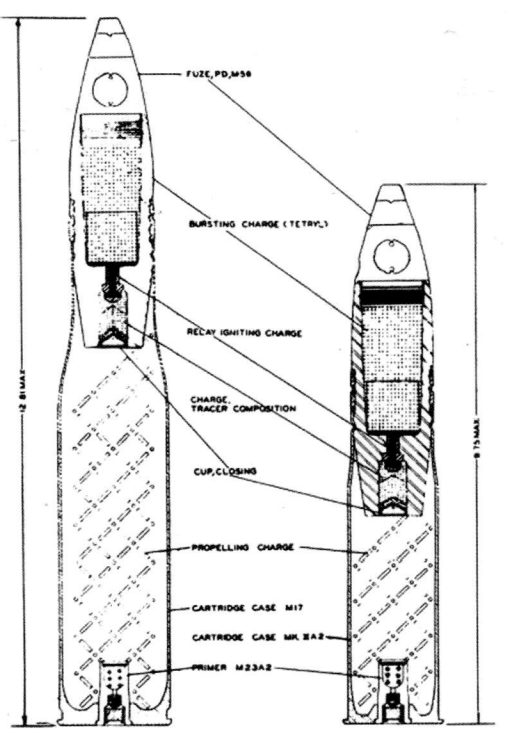

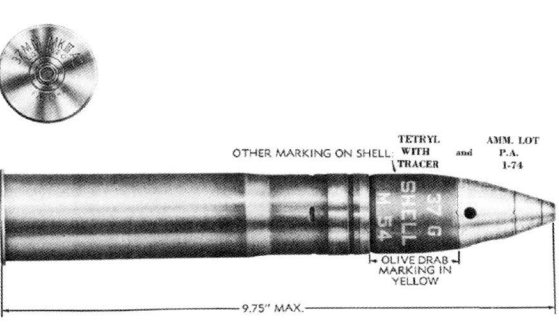

The M10 was an improved version of the M4 and fitted to the Kingcobra, an improved version of the Airacobra. Color codes used: yellow (later in the war olive drab with yellow lettering) is high explosive, red is low explosive and shrapnel, solid kinetic rounds are painted black, practice ammunition is painted blue, and chemical rounds have a blue-gray color with circumferential bands painted on the round representing the type of chemical, such as white phosphorus smoke.

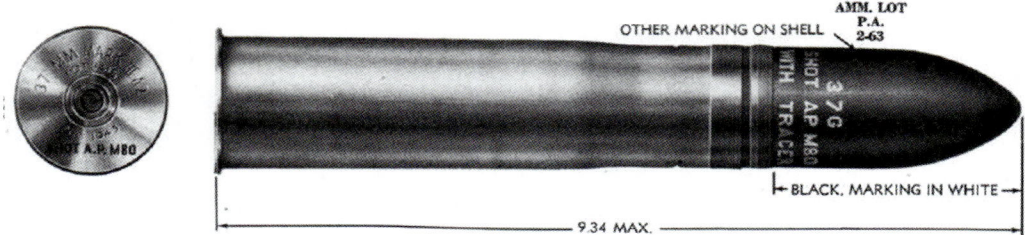

Armor-piercing cartridges M80 for 37-mm aircraft guns AN-M4 and M10. (TM 9-240)

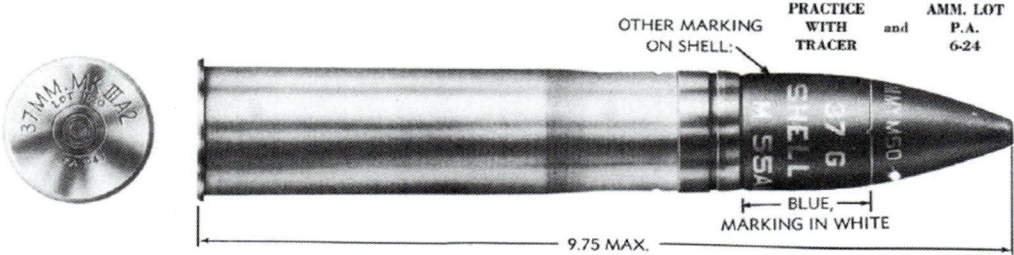

Practice cartridge M55A1 for 37-mm aircraft guns AN-M4 and M10. (TM 9-240)

Cartridge, fixed, Practice M55A1 w/tracer. (Roberts Collection)

Cartridge, Fixed, Practice M55A1 With Tracer

This round was used to simulate firing the M54 HE cartridge. The cartridge case of the M55A1 was the M17 brass case or the M17B1 steel case for the M1A2 gun. The M4 gun used the cartridge case Mk IIIA2. The standard M23A2, 20-grain, percussion primer was used. The propellant was 6 ounces of FNH powder. The projectile is the same as the HE projectile M54 with the exception of explosive filler. The fuze, M50, was a dummy and made of aluminum. The weight of the round was the same as the M54. The round was 12.81 inches in length and weighed 2.62 lb.

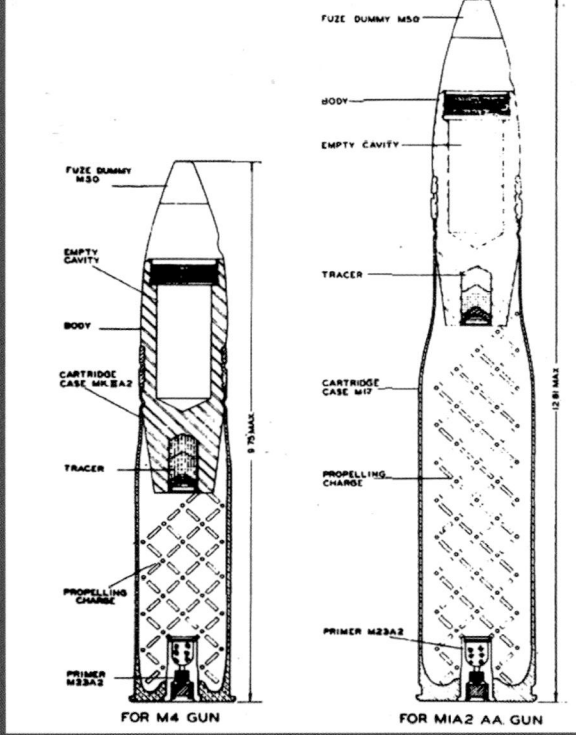

Cartridge, fixed, Practice M55A1 w/ tracer for 37-mm guns M1A2 and M4. (TM 9-1904)

Cartridge, Fixed, APC M59 With Tracer

The APC M59 is designed to defeat any type of armor plate material providing the thickness is within the capability of the round. The cartridge case is the standard brass M17 and the steel M17B1. The primer is the M23A2, 20-grain, percussion primer which is standard for these rounds. The propellant charge is 0.31 lb of FNH powder yielding a muzzle velocity of 2050 ft/sec. The projectile is similar to the APC M51 used in tank and antitank guns M3, M3A1, M5 and M6. The complete round is 12.76 inches long and weighs 3.12 lb. The projectile has a soft metal cap that tends to grab the armor plate when striking at an angle, reducing the chance of a ricochet. The main difference is the M59 has a lower-caliber ogive than the M51 because the projectile does not have a streamlining nose cone.

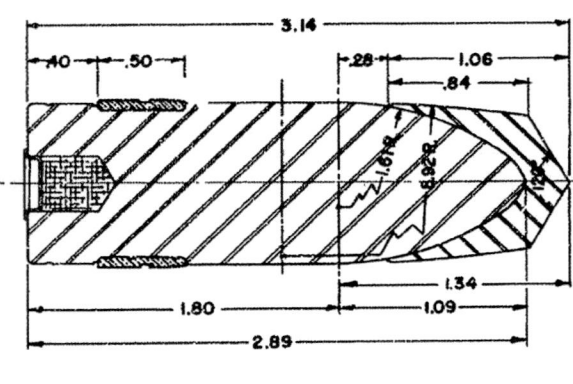

Projectile APC M59 w/tracer (Handbook of Ballistic and Engineering Data for Ammunition, Vol. 1)

Cartridge, Fixed, AP, M74 With Tracer

This projectile is effective against homogeneous armor plate, but is not as effective against face-hardened armor. The cartridge case is either brass M16 or steel M16B1 for the tank and antitank guns and M17 or M17B1 for the M1A1 gun. The standard M23A2, 20-grain primer is used. The cartridge case contains approximately 4 ounces of FNH powder yielding a muzzle velocity of 2,900 ft/sec. The projectile is the same as the M74 used in the 37-mm tank and antitank guns M5, M6, M3, and M3A1. The complete round is 13.01 inches long, weighs 3.07 lb with a 1.5-caliber ogive—a rather stubby projectile

37-mm M74 cartridge. (Roberts Collection)

37-mm Automatic Gun M4 (Aircraft)

This gun, designed for aircraft, has a typical muzzle velocity of approximately 2,000 ft/sec. The ammunition is supplied to the gun by a 5-round feeder clip, by a 15-round disintegrating belt, or by a 30-round M6 endless belt. The cartridge case has an extracting flange but no grooves. The types of ammunition used are high-explosive (tetryl), armor-piercing (kinetic), and practice.

The Mark IIIA2 cartridge case is standard for all ammunition used in the M4 gun which is 5.69 inches long with a flange for the extractor mechanism. The M23A2, 20-grain, percussion primer was standard for all aircraft cartridges. The propellant was 2.5 ounces of FNH powder for the high-explosive and practice rounds. The armor-piercing rounds required 2.3 ounces of FNH powder. The projectile is the same as the M54, HE projectile used in the M1A2 antiaircraft gun. The round is 9.75 inches in length and weighs 1.94 lb. The practice shell, M55A1 w/ tracer, is the same as the M55A1 practice shell for the M1A2 antiaircraft gun.

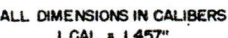

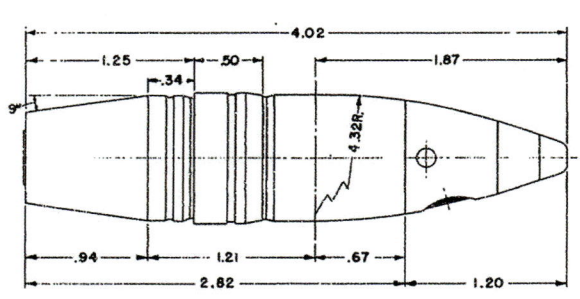

ALL DIMENSIONS IN CALIBERS
I CAL = 1.457"

SHELL, HE, 37-MM, M54
FUZE, PD, M56

Projectile 37-mm M54. This is a typical example of an approximately 5-caliber-ogive projectile without the drive band detail. These drawing are typically dimensioned in calibers. In this case, multiply all measurements by 1.457 inches/caliber to obtain the dimensions in inches. (Handbook of Ballistic and Engineering Data for Ammunition, Vol. 1)

Drill cartridge M23 for 37-mm gun M4. (Roberts Collection)

Cartridge, Fixed, Armor-piercing, M80 With Tracer

This cartridge, which is fired from the M4 and M10 aircraft gun, is similar to the M59 without the piercing cap. The M80 is lighter in weight because of the shortened projectile. The M80 projectile is 4.23 inches in length and weighs 1.66 lb. The radius of ogive is 2.35 inches. The complete round is 9.34 inches long and weighs 2.25 lb. The cartridge case has an extraction flange and no groove.

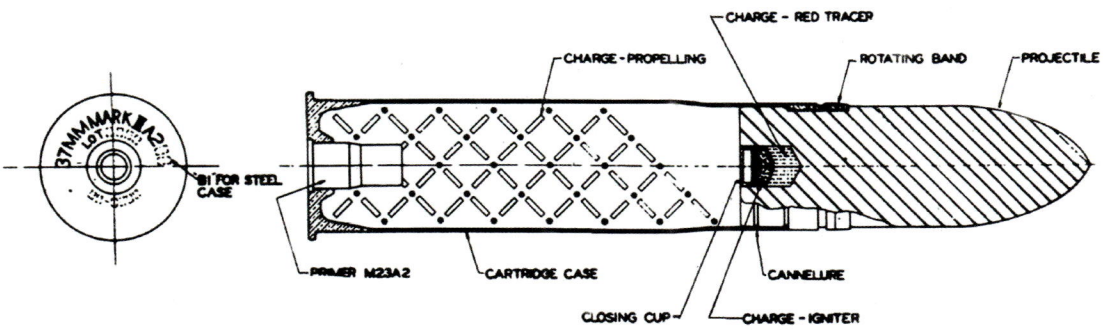

M80 37-mm gun M4 cartridge. (TM 9-1904)

37-mm M80 projectile, manufactured as a part of lot number E-E-6-11-1944. (Roberts Collection)

37-mm Gun T32

There were two types of ammunition authorized to be used by the 37-mm gun T32. They were cartridge HE M63 with base-detonating fuze M58, and canister M2.

The Mk IIIA2 cartridge case, used with the 37-mm gun T32, was filled with FNH propellant yielding a muzzle velocity of 1,500 ft/sec. The cartridge case was fitted with primer, percussion, 55-grain M38A1 or M38B2. The HE round weighed approximately 2.25 lb with a length of 10.95 inches. The weight of the projectile was 1.61 lb filled with 0.085 lb of TNT. The fuze was base detonating, non-delay. The weight of the canister round was 2.25 lb with a length of 10.95 inches. The filler was approximately 122 steel balls in a matrix and weighed approximately 1.22 lb.

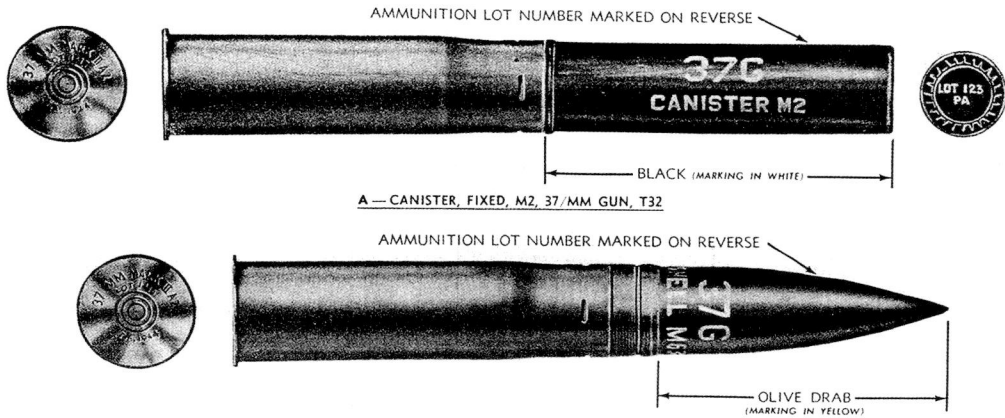

Ammunition for the 37-mm gun T32. (TM 9-246)

Squeeze Bore 37-mm Projectile

The squeeze bore projectile was invented by a German, Karl Puff, in 1903. The basic principle is that a round fired through a tapered bore will squeeze down in diameter, thereby increasing the velocity of the projectile. This worked well, but the high pressure and wear from squeezing the soft metal body took a toll on the gun tube, reducing its useful life by as much as 80%. This was a feature on the M22 light tank described following.

Cartridge, Canister, Fixed M2

This round is used against personnel and is essentially a large shotgun shell. The cartridge case is the M16 and is standard for this round. The M23A2, 20-grain, percussion primer is standard. The propellant charge is 0.44 lb of FNH powder loosely packed into the cartridge, resulting in a muzzle velocity of 2,500 ft/sec. The canister case is a thin-walled cylinder containing 122 steel balls of 0.375 inches in diameter imbedded in a resin material. The canister is 6.50 inches long and weighs 1.9 lb. The cartridge case is 8.75 inches long with extracting flange only. The complete round is 14.53 inches long and weighs 3.31 lb.

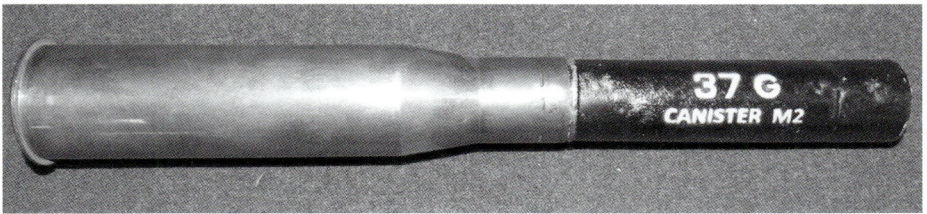

Canister anti-personnel 37-mm M2 cartridge. This is essentially a shotgun shell with steel shot scattering in a pattern, making it deadly for personnel. (Roberts Collection)

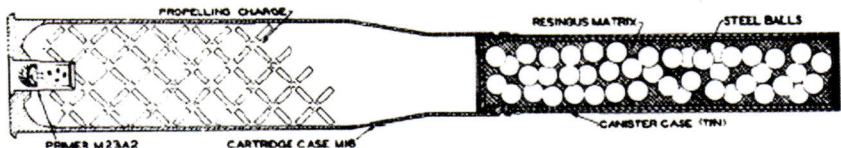

Canister cartridge M2. (TM 9-1904)

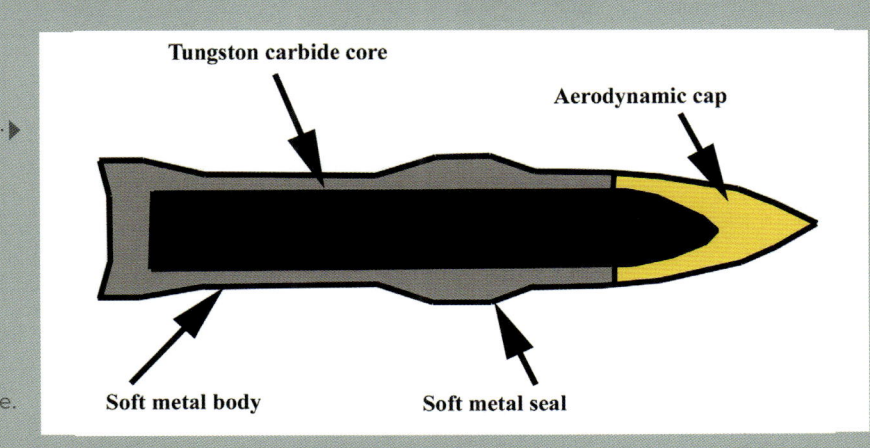

Squeeze bore projectile. (Roberts Collection)

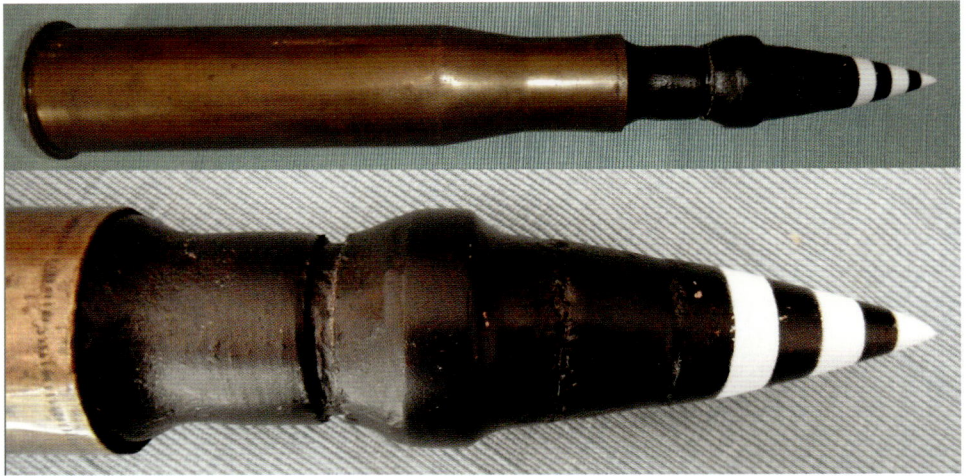

37-mm Littlejohn projectile. (Jeff Wszolek private collection)

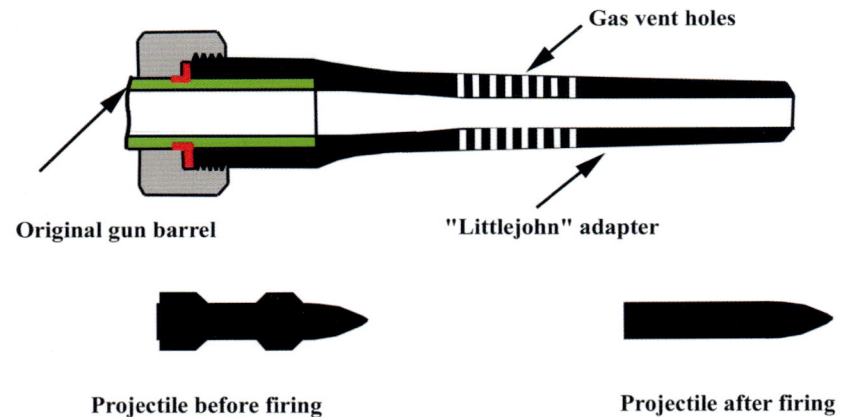

Gas vent holes

Original gun barrel

"Littlejohn" adapter

Projectile before firing

Projectile after firing

The projectile was tungsten carbide with an outer, soft metal casing (aluminum alloy). As it passed through the squeeze bore adapter, the diameter reduced to approximately 30 mm and the muzzle velocity increased to about 4,000 ft/sec. Tungsten carbide is approximately twice the weight of steel per cubic inch which adds significantly to the penetrating capability. (C. Roberts, *U.S. Airborne Tanks 1939–1945*)

M22 Locust light tank with Littlejohn squeeze bore adapter on a 37-mm gun M6. A drawing of the squeeze bore adapter is shown on the previous page. (C. Roberts, *U.S. Airborne Tanks 1939–1945*)

The Ordnance Department at Aberdeen Proving Grounds in Maryland experimented with a squeeze bore adapter for the M3 37-mm gun in early 1942. The purpose was to increase the penetration capability of the 37-mm gun. The British Littlejohn adapter was also tested. Rather than design a tapered barrel, the standard 37-mm barrel was threaded at the end to accept a connection collar and the squeeze bore adapter. The projectile flanges were constructed of a soft metal that formed a seal that prevented propellant gases from escaping. The cupped shape at the base of the projectile allowed the propellant gas pressure to push the flanges against the barrel.

The base flange was essentially the drive band; the forward flange was essentially the bourrelet. The squeeze bore concept was mostly used by the Germans with the help of engineer, Hermann Gerlich, but many countries attempted to copy the design, often without success. Gun barrel life was significantly reduced because of the increased wear from the projectile. Firing chamber pressure was high. The space between the base and forward flanges would often entrap propellant gas. When the round was squeezed, this space would decrease, causing excessively high pressure and damage to the flanges as the round exited the barrel. The damaged flanges would separate from the projectile upon exiting the muzzle. Gas vent holes were drilled in the forward flange to help relieve this pressure, but with the sudden pressure rise, this did not solve the problem. There was increased

Rare photo of the 37-mm gun and the squeeze bore extension and connection collar T1, 3/28/1942. (U.S. Army)

Rare photo of the 37-mm gun with squeeze bore adapter T1 mounted. Unlike the British design, the adapter came in two pieces, the squeeze bore adapter and connection collar. In the British design the connection threading was integral to the adapter and required only one piece, the squeeze bore adapter, 3/28/1942. (U.S. Army)

deceleration of the projectile, resulting in diminished effectiveness at long ranges which was probably a result of the folded flanges. No doubt the researchers at Aberdeen encountered similar problems. Placing an adapter on a normal gun may have been a problem. When the projectile accelerates after being fired, it impacts the adapter and starts to squeeze. It is not clear how much extra stress this puts on the gun tube. A decision was made at Aberdeen not to develop such a weapon for the United States. The Germans stopped using these guns because of a shortage of tungsten.

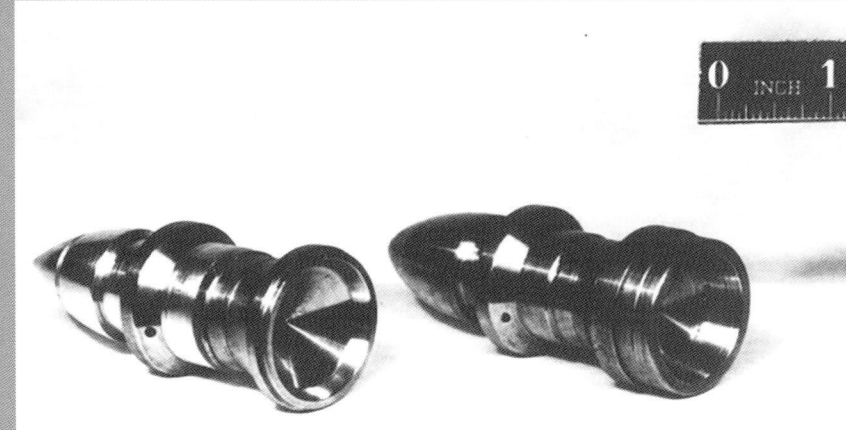

Bottom view of squeeze bore projectiles. High velocity 37/28-mm AP shot T23 on the left; on the right is projectile, special, 37/28-mm, copper body, 5/1/1942. (U.S. Army)

High-velocity 37/28-mm AP shot T23 on the left; on the right is projectile, special, 37/28-mm, copper body, 5/1/1942. The shot started out as 37 mm and was reduced to 28 mm when exiting the gun. (U.S. Army)

Spigot Grenade 37-mm Antitank Gun, T21 & T22

One project at the Aberdeen Maryland Proving Grounds was to develop a weapon that was effective against Japanese troops as well as Japanese fortifications. The researchers constructed spigot grenades based on the German design. It looked like they could be adapted to the M3 37-mm antitank gun. The grenade T21 was developed as an antiarmor or concrete device. This device weighed approximately 7.88 lb and was equipped with a shaped-charge warhead using cyclotol as the explosive. Cyclotol is a combination of Research Department Explosive (RDX) $(O_2N_2CH_2)_3$ and TNT. When fired, the velocity of the grenade was approximately 489 ft/sec. The round could penetrate approximately 10 inches of homogeneous armor plate. Testing by the Marine Corps indicated a maximum effective range of the T21 of approximately 600 yards and 500 yards for the T22. The Marine Corps reported that the spigot grenade would be effective against Japanese bunkers and other strongpoints. The Marine Corps Equipment Board report on the performance of this munition was favorable but concluded that it would be too late as the war had just ended.

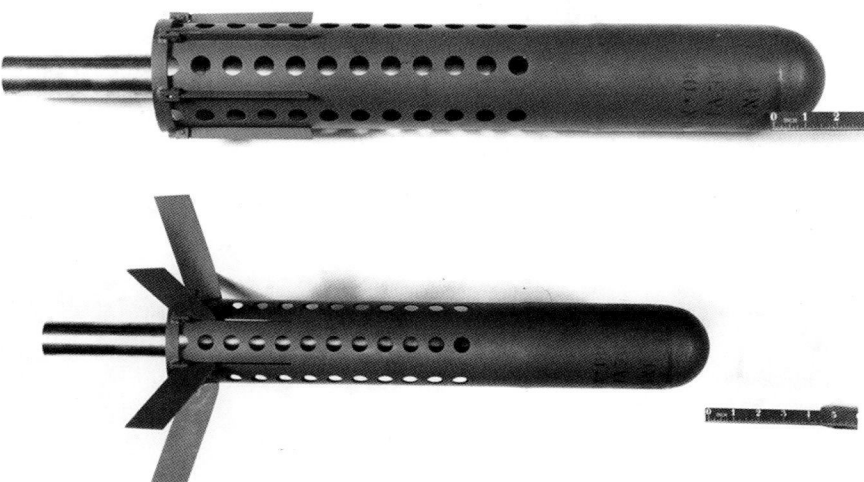

Spigot grenade T22 with tailfins folded (upper unit in the photo) and tailfins deployed (lower unit in the photos), May 1945. (U.S. Army)

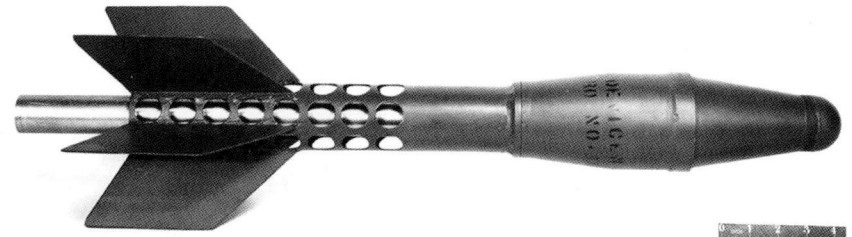

Spigot grenade T21 tested at Aberdeen Proving Grounds, March 1945. (U.S. Army)

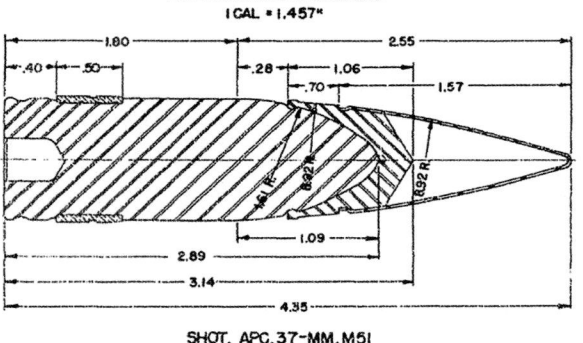

Rare photo of spigot grenade T21 loaded in a 37-mm M3. (U.S. Army)

Performance of the M51 Cartridge

There are three criteria for armor penetration: Army, Navy, and protection. To satisfy the Army criterion, the target must be breeched so that light shines through the penetration. The Navy criterion states that at least half the projectile passes through the target. The protection criterion stipulates that a thin aluminum sheet behind the target must be perforated by impact fragments to be considered penetrated. In the graph on the next page, for a projectile velocity of 2,800 ft/sec striking a homogeneous steel plate at 30° angle, the penetration is approximately 2.4 inches.

ALL DIMENSIONS IN CALIBERS
I CAL = 1.457"

SHOT, APC, 37-MM, M51

Drawing of the M51 projectile. The dimensions are in calibers. Multiply each dimension by 1.457 to obtain the dimension in inches. (Handbook of Ballistic and Engineering Data for Ammunition, Vol. 1)

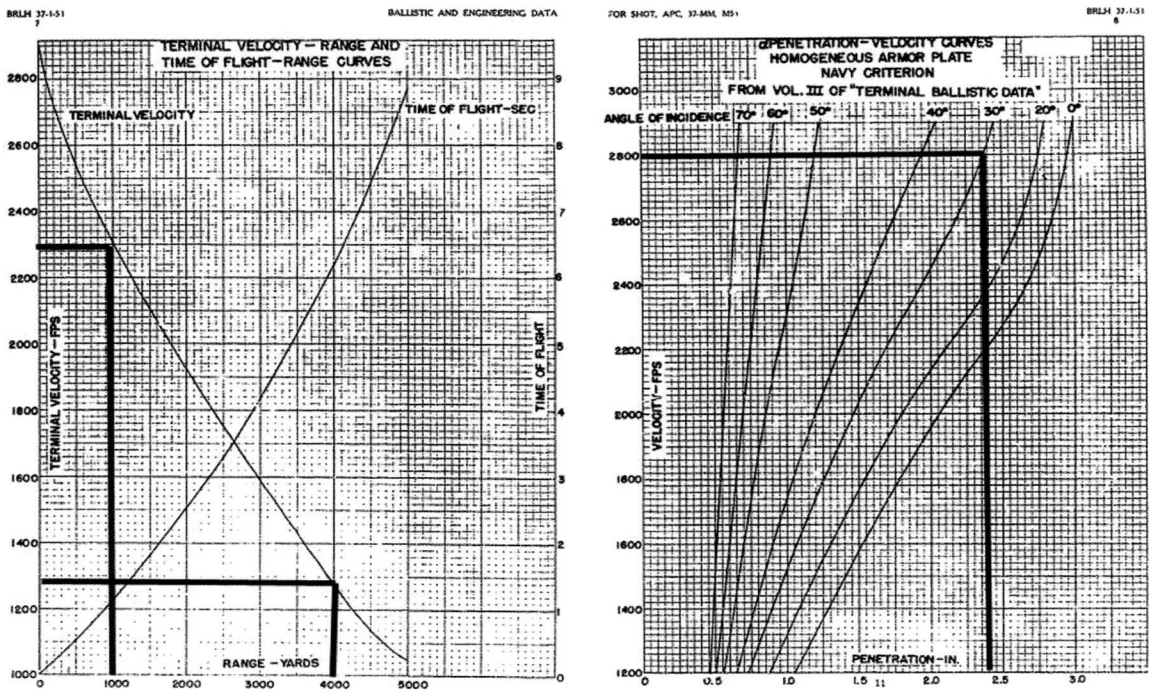

(Left) Terminal velocity and time of flight data for the APC M51 cartridge in the 37-mm gun. At 4,000 yards, the projectile velocity is approximately 1,280 ft/sec. At 1,000 yards the projectile velocity is 2,300 ft/sec. (Handbook of Ballistic and Engineering Data for Ammunition, Vol. 1)

(Right) Penetration velocity curves for APC shot 37-mm M51 against homogeneous armor plate using Navy criterion. There are three criteria for armor penetration: Army, Navy, and protection. To satisfy the Army criterion, the target must be breeched so that light shines through the penetration. The Navy criterion states, that at least half the projectile passes through the target. The protection criterion stipulates that a thin aluminum sheet behind the target must be perforated by impact fragments to be considered penetrated. In this graph, for a projectile velocity of 2,800 ft/sec striking a homogeneous steel plate at 30° angle, the penetration is approximately 2.4 inches.

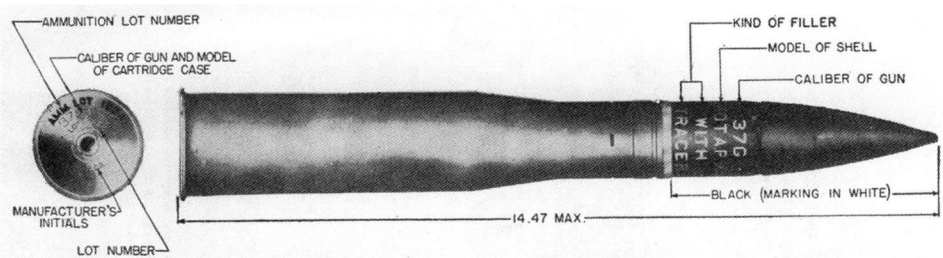

Armor-piercing, capped cartridge M51. (FM 23-81)

M2A4 light tank, 70th Tank Battalion November 1941. (C. Roberts, *Armored Strike Force: The Photo History of the 70th Tank Battalion in World War II*)

| 4 37-mm Gun With Armored Vehicles in World War II

The M2A4 was the first tank entering service to be equipped with the 37-mm gun M5. Production started in May 1940 at the American Car and Foundry Company and continued through March 1941 with a production run of 375 tanks. It had 1-inch-thick armor and a seven-cylinder Continental W-670 radial aircraft engine.

M2A4 Light Tank

The U.S. Army considered the M2A4 to be an interim design which led the way to the development and production of the M3 light tank. Consequently, most of the M2A4 tanks were relegated to training between 1940 and 1942. Approximately 36 M2A4s were delivered to the Marine Corps, who used them during the Guadalcanal landings. M2A4s and M3 Stuarts were assigned to Marine Corps units using them to knock out Japanese bunkers, strongpoints, and tanks. Canister shot, 37-mm, was used against Japanese infantry. During operations, two tanks would park back-to-back to guard each other from attacks by Japanese infantry with hand-carried charges. In early 1941, the British received 36 M2A4s but favored the much-improved M3 Stuart which served in North Africa. After 1943, M2A4s were used for training.

M2A4 light tank, Guadalcanal. (Wikimedia Commons)

There were four .30 cal. machine guns in the M2A4, one in the turret, one on the bow and one in each of the sponsons left and right. The main gun was the 37-mm M5 with manually operated breech. There was no turret basket, so the commander/loader stood on the right while the gunner stood on the left. The commander/loader turned the turret in the direction of the target and the gunner would use the M5, telescopic sight to aim at the target. The gun mount, M20, had 20° of traverse capability and elevation/depression capability. It should be noted that this vehicle's undercarriage was the basis for the design of all the Stuart variants that would appear in the future.

M3 Stuart Light Tank

After observing events in the Far East and in Europe, it became apparent that the M2A4 light tanks needed improvements. The Ordnance Department approved a new design with thicker armor, an upgraded suspension system, and new gun recoil system, designating it the M3 light tank. The final design included the slightly longer 37-mm M6 gun and machine guns in the turret, on the bow and on an antiaircraft mount on the turret. It was equipped with 38 mm armor on the upper front, 44 mm on the lower front, 51 mm on the mantlet, 38 mm on the turret sides, 25 mm on the hull sides, and 25 mm on the rear hull. Approximately 8,936 tanks were built with the seven-cylinder Continental radial gasoline engine (W-670) or the Guiberson nine-cylinder T-1020 radial diesel engine. The M3 tanks, like the M2A4s, had a very high profile, since the radial engine had to be mounted with the crankshaft horizontally facing the front, making them easy to spot by the enemy at a distance. The M3 had the rear idler lowered to make ground contact to support the additional weight of the tank when compared to the M2A4. The riveted hull was deficient in that if an enemy projectile struck and sheared off a rivet head, the rivet head inside the tank would ricochet around the fighting compartment, causing injury to the crew

The British were the first to use the M3 Stuart in combat in the North African campaign with a total force of over 700 tanks. The M3 was comparable to small and medium German tanks and superior to most Italian tanks. However, losses of the M3 were high, mostly due to better tactics by the Germans. The British reporting on the performance of the M3 was a mixed bag. The M3 Stuart with the 37-mm M5 gun would not defeat the larger German tanks due to a poor internal layout, two-man turret crew (three preferred), and limited range. On the positive side, the crews liked the high speed, reliability of the tank, and the use of high-explosive shells. The crews often referred to the M3 as "Honey." By mid-1942, the British relegated the Stuarts to non-tank combat roles such as reconnaissance and as auxiliary transport vehicles with the turrets removed.

53183   11-17-41   ABERDEEN PROVING GROUND   ORDNANCE DEPT.
Project No. 2-7. Light Tank M3, No. 1946 (A. C. F.). Left view. NOTE Homogeneous Plate Turret and Pistol Port -- Storage Box on rear fender.

M3 Stuart light tank. (Roberts Collection)

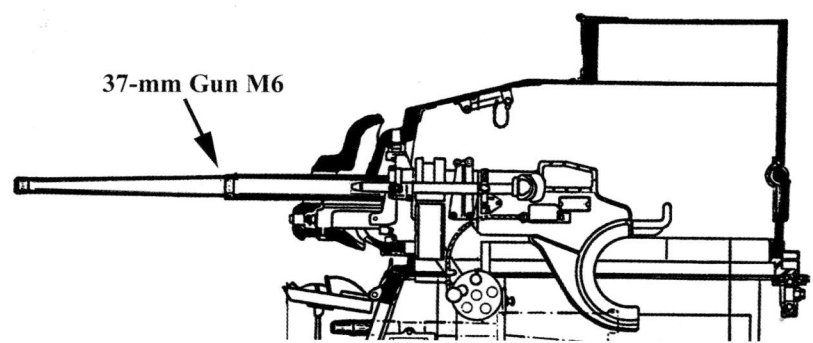

M6 gun elevation drawing. (TM 9-726)

The Russians received M3s under Lend-Lease but were dissatisfied with their performance. The M3's radial aircraft engine required high-octane fuel which was practically unavailable in the Soviet Union. The Russians considered the tank to be under-gunned, under-armored and sensitive to fire development when hit. Despite this, the M3 tank was used until the end of the war.

The U.S. Army sent 108 M3 Stuart tanks to the Philippines with the 194th and 192nd Tank Battalions. These tanks were eventually given up in the surrender to the Japanese and were used by the Japanese. The Marines successfully used the M3 in jungle warfare but transitioned to medium tanks by the end of the war. In North Africa, the U.S. Army used M3 Stuarts but relied on the M4 Shermans for the tank-to-tank confrontations. However, in Sicily, the 70th Tank Battalion, using M5 Stuarts, engaged a column of German Mark IIIs by attacking from hilltops, turning them back. The 37-mm gun was effective against the thinner German top armor which was vulnerable to a top attack. In Europe, the M3 was used for reconnaissance and auxiliary carrier roles with the turret removed. The M3 was also used by France, China, and Yugoslavia. There were several variants of the M3 Stuart including British M3 (Stuart I), M3 diesel (Stuart II), M3A1 (Stuart III), M3A1 diesel (Stuart IV), and M3A3 (Stuart V).

M5 Light Tank

The M5A1 Stuart was the final version of the M2 tank series. It incorporated two Cadillac engines to replace the radial aircraft engines which were in high demand. Improvements included Hydra-Matic transmissions, making driver training much easier than clutch-application vehicles, and a gyroscope-driven gun stabilizer for the M6 37-mm gun, making it easier to aim.

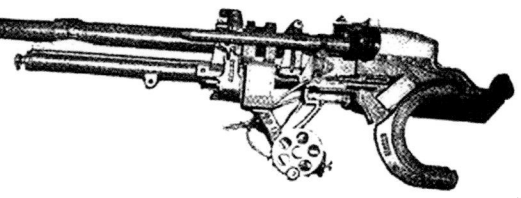

M6 gun mount for light tank. (FM 23-81)

Although the 37-mm M6 was ineffective against later Panzer IV, Panzer V (frontal armor), and Panzer VI tanks, it was effective against all Japanese and Italian armor, German halftracks and armored cars, Panzer I tanks, Panzer II tanks, and early Panzer III tanks. The HE rounds were not powerful enough to support infantry in a majority of the engagements but were effective against lightly armored vehicles. The canister rounds were effective against Japanese infantry in the Far East.

In Profile:
T33 37-mm Gun Motor Carriage & Ford 1½-ton

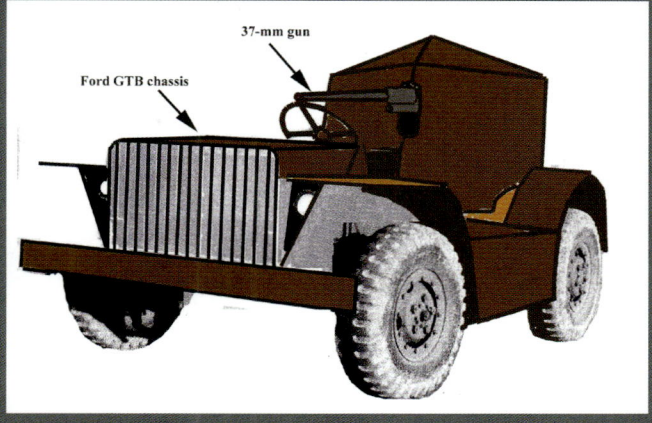

T33 37-mm Gun Motor Carriage. (Roberts Collection)

Ford 1½-ton GTBA Truck, "Burma Jeep." (Roberts Collection)

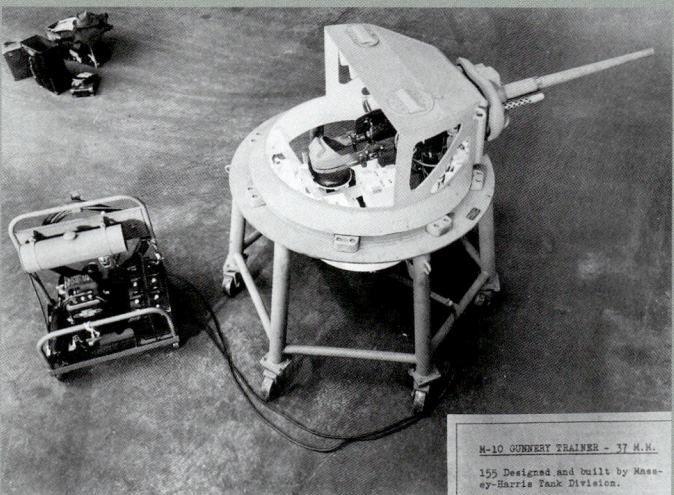

Restored M5A1 Stuart light tank in a reenactment. (Roberts Collection)

M-10 GUNNERY TRAINER - 37 M.M.

155 Designed and built by Massey-Harris Tank Division.

With the development of the M5 Light Tank equipped with the gun stabilizing system, there was a need for training on how to use this system. Approximately 154 M-10 37-mm Gunnery Trainers were produced and issued to tank units and training centers. Tank crews would learn and practice how to use the gun tube leveling system and other controls associated with the turret. Crews and maintenance personnel would train on how to test, calibrate, maintain and repair the system. The trainer was equipped with an independent 12 volt power unit consisting of a Briggs and Stratton gasoline fueled generator and 12 volt batteries. (U.S. Army)

LVT-(A)1

The LVT (landing vehicle tracked) was an amphibious assault vehicle designed to pass over reefs and run up the beach, unloading troops. It was found that these vehicles needed more firepower than machine guns when landing on a beach. Consequently, the vehicle was up-armored, and an M5 tank turret with a 37-mm M6 gun in an M44 mount was added. The vehicle was designated LVT-(A)1, where A stands for armored. The vehicle armor was from 6 mm to 12 mm thick and the power source was a 262 brake horsepower, air-cooled, radial engine. The purpose of the vehicle was to provide artillery support at the beginning of Marine landings. The 37-mm gun was moderately successful but was eventually replaced with the 75-mm gun in the LVT-(A)4 vehicles.

(Above) LVT-(A)1 right front view. (TM 9-775)

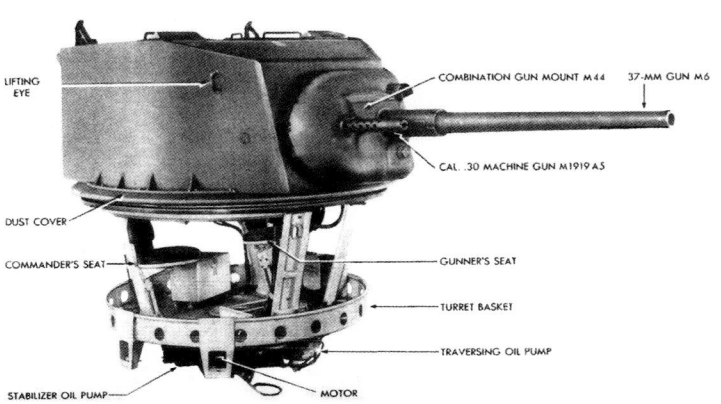

(Left) LVT-(A)1 turret, the same as the M5 tank turret. (TM 9-775)

LVT-(A)1 upper deck.
(Roberts Collection)

M6 37-mm gun and
stabilizer. (TM 9-775)

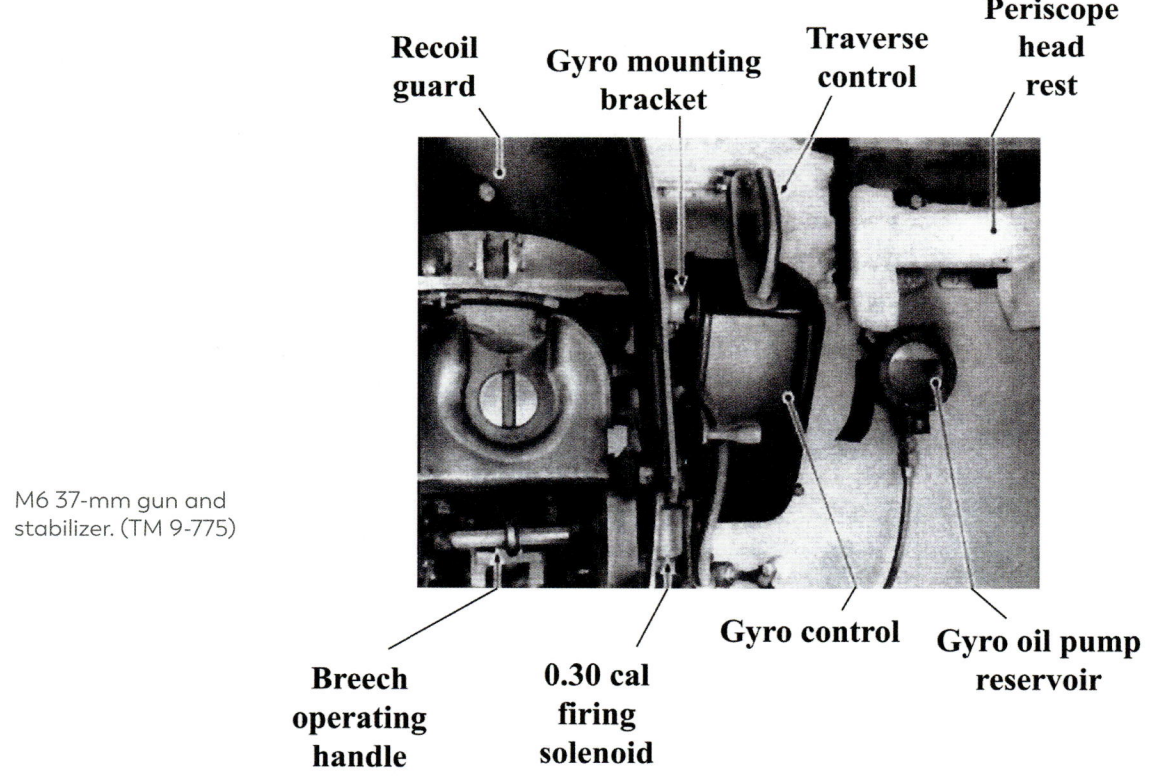

M8 Armored Car (Greyhound)

In mid-1941, the Ordnance Department started development of a fast tank destroyer to replace the M6 37-mm gun motor carriage (see Chapter 5). Specifications included a 6x6 wheeled vehicle, a turret with a 37-mm gun, coaxial .30-cal machine gun, a mount for a .50-cal antiaircraft gun, frontal armor to defeat .50-cal projectiles, and side armor to defeat .30-cal projectiles. Prototypes were submitted by Studebaker (T21), Ford (T22) and Chrysler (T23). The Ford Motor Car Company was selected to manufacture the M8.

M8 armored car in Europe. (Roberts Collection)

As production started, it was clear that the 37-mm gun M6 could not act as a tank destroyer. There was also disagreement amongst Armored Force Board members as to the need of this vehicle. The M8 was removed from tank-destroyer assignment to reconnaissance assignments. Production started in March 1943 and ended in June 1945 with over 8,500 vehicles manufactured by Ford. Ford plants in Chicago and St. Paul, Minnesota manufactured the vehicles. The British Tank Mission turned down the M8 Greyhound (Lend-Lease), the T18 Boarhound, and the M38 Wolfhound. The British did adopt the T17E1 Staghound. (Eventually 1,000 Staghound vehicles were given under Lend-Lease to the United Kingdom, France, and to Brazil.) The M8 was fast, a plus for reconnaissance units.

The M8 was often used to destroy soft targets with the 37-mm gun and quickly leave the scene. In the reconnaissance role, the M8 was equipped with long-range and short-range radios to report on what was ahead of the main unit. The M8 weighed 17,400 lb, had a range of 100–200 miles over land and 200–400 miles on paved roads with a speed capability of up to 55 mph. The 37-mm gun M6 worked well in this reconnaissance role.

M8 armored car left front view. (Roberts Collection)

M8 armored car right view. (TM 9-743)

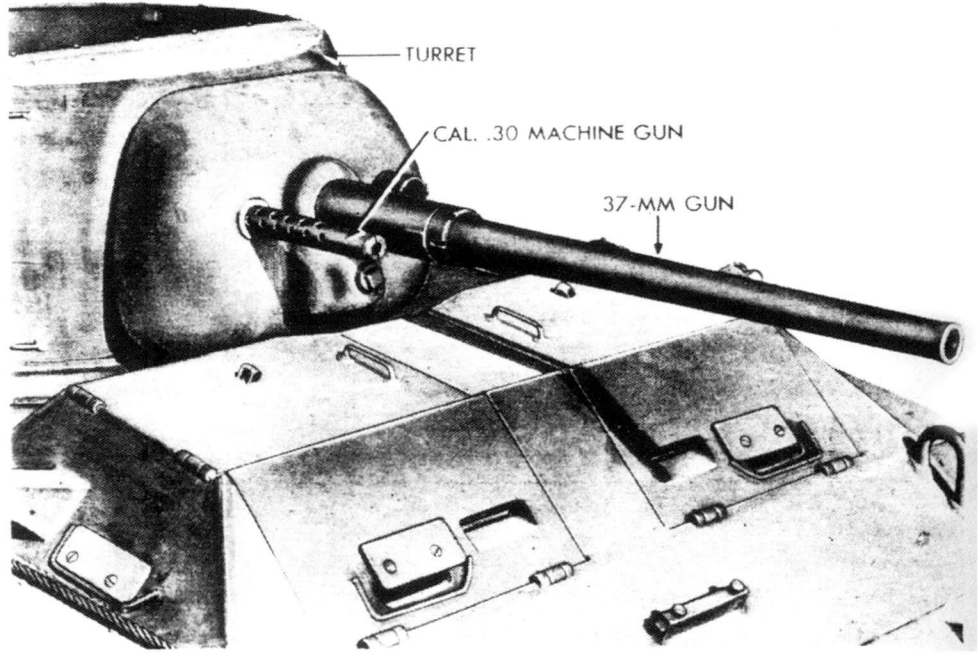

Right front view of M8 showing .30-cal coaxial machine gun and 37-mm M6 main gun. (TM 9-743)

M22 Locust light tank in Germany, circa spring 1945. George Moodie is the driver. (C. Roberts, *U.S. Airborne Tanks 1939–1945*)

M22 "Locust" Airborne Tank

The M22 airborne tank was the only American-designed tank delivered to a battlefield by glider in World War II, during Operation *Varsity*. It was designed in response to the British desire to have a light tank to accompany airborne troops. The M22 was equipped with a 37-mm M6 gun and coaxial .30-cal machine gun. The doctrine was to have some armor for the lightly armed airborne troops so they could hold positions until the heavier tanks arrived. To save weight, gun leveling and turret traversing systems were eliminated in favor of a manual system. During Operation *Varsity*, it was able to do its job holding off German infantry.

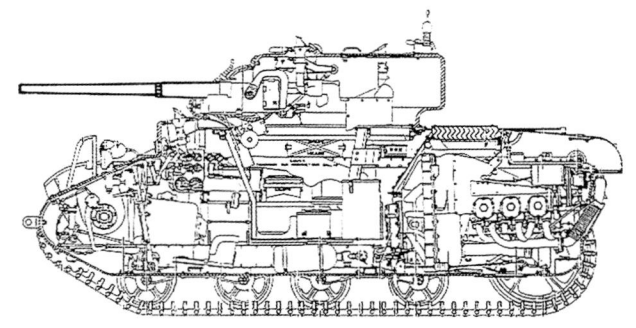

Drawing of the M22 Locust airborne tank showing the 37-mm gun M6. (C. Roberts, *U.S. Airborne Tanks 1939–1945*)

M3 "Lee/Grant" Medium Tank

The Medium Tank M3 "Lee" was the first American medium tank used in World War II. The British variant used a different turret and was referred to as the "Grant." The design of the M3 medium tank was developed in July 1940, and the first versions were available in late 1941.

The U.S. Army needed a tank with a 75-mm gun, and the British badly needed such a tank for the North African campaign, so the M3 was a compromise, giving speed of its development a high priority. The M3 Lee was equipped with a 75-mm main gun in a sponson, a 37-mm M5 gun on a turret and machine gun in a cupola. Problems with this design were the high silhouette to accommodate the large aircraft radial engine, a sponson-mounted 75-mm gun requiring the tank to be moved in order to bring the target to bear, riveted construction and poor off-road performance. Riveted construction was a problem in that when an enemy projectile struck a rivet and knocked the head off, the other end of the rivet would ricochet around the fighting compartment, injuring the crew. This is because the rivets are under high tension. Another problem with the radial engine is that you must crank it over by hand several turns in order to scavenge the residual oil in the bottom cylinder. If this was not done, the bottom cylinder connecting rod could be damaged as a result of hydrostatic lock. Despite these deficiencies it held its own against the German tanks early in the North African campaign. As soon as the M4 Sherman was available, the M3 Lee was taken out of combat operations and used in other roles such as the T2 tank retriever.

The British used the M3 Grant tank in the Far East, and M3 Lee tanks were given to the Soviets, under Lend-Lease. The 37-mm gun used an M2 periscope mounted in the mantlet, which also sighted the coaxial machine gun. The range scale provided up to a 1,500-yard range for the 37-mm gun. In the movie *Sahara*, with Humphrey Bogart, the Lee tank was portrayed shooting down a German aircraft with the 37-mm gun.

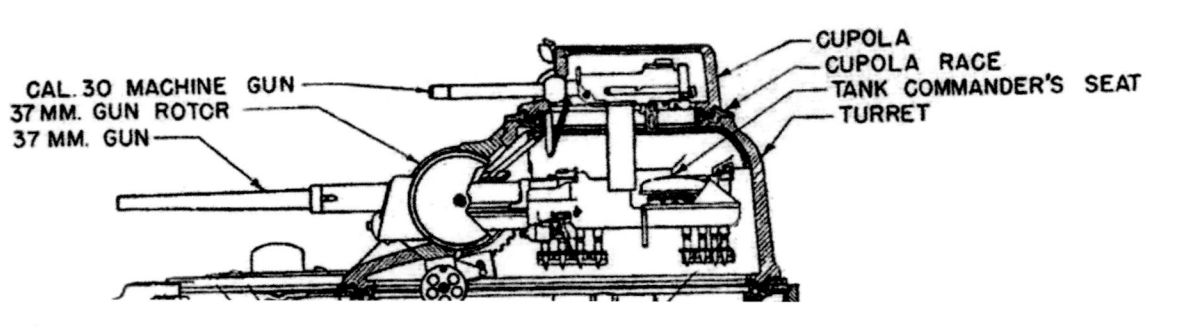

M3 Lee medium tank turret detail. (TM 9-750)

M3 Medium Tank "Grant." (Roberts Collection)

Comparison of the Grant tank on the left and Lee tank on the right. The British used mostly Grant tanks, but there were some Lee tanks used. The U.S. Army used Lee tanks but no Grant tanks. The British eliminated the machine gun cupola and added more room inside the turret for radio equipment to form the Grant tank. (Wikimedia Commons)

T17E1 Staghound Armored Car

In July 1941, the Ordnance Department published specifications for a medium armored car (T17E1 Staghound) and a heavy armored car (T18 Boarhound). The British Purchasing Commission eventually adopted the T17E1 Staghound armored car. Over 3,800 Staghounds were manufactured by Chevrolet and sent to the British. The Staghound was used in Europe, mostly by cavalry squadrons. It was equipped with the M6 gun. The U.S. Armored Board convened and decided to settle on the M8 Greyhound as the standard armored car for the American forces.

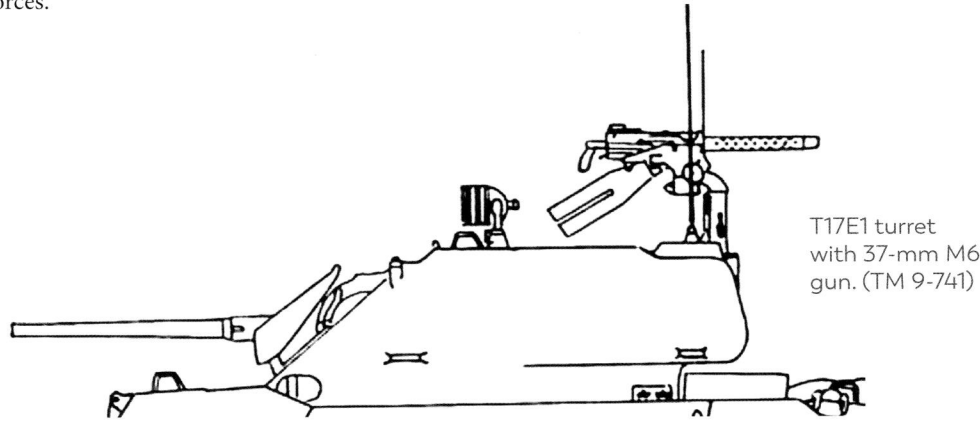

T17E1 turret with 37-mm M6 gun. (TM 9-741)

T17E1 Staghound armored car. (TM 9-741)

Scout Car M3A1E3 with 37-mm gun. (Wikimedia Commons)

Scout Car M3A1E3 and Halftrack with 37-mm Gun

In June 1940, the Cavalry Department came up with the idea to mount a 37-mm antitank gun on the M3A1 scout car that was already in production and issued to cavalry units. The 37-mm antitank gun on a T6 gun mount was placed just ahead of the rear axle. The unit tested very well at the Aberdeen Proving Grounds, but some deficiencies were noted. First, it was thought that the vehicle had too high a profile with the 37-mm mounted. Second, it was considered inefficient to have such a large vehicle carrying only a 37-mm gun, which was becoming obsolete. The fact that the M6 GMC was in production negated the need for this vehicle.

Aberdeen Proving Grounds canceled the project but sent the test vehicle to Camp Funston, Kansas, for further evaluation by the Cavalry Board. The Cavalry Board made modifications to the gun and mount but eventually canceled the project.

37-mm antitank gun M3 mounted on a Scout Car M3A1E3 using the modified T-6 mount. The T-6 modified mount is about 17 inches farther back than the original T-6 mount. The soldier has the 2nd Cavalry Division patch and the location is most likely Camp Funston, Kansas. A flash suppressor is mounted at the muzzle. (Roberts Collection)

The same soldier aiming the gun aft. (Roberts Collection)

This halftrack was armed by simply placing an M3 37-mm gun in the cargo compartment and securing it to the chassis. This is a typical early war adaptation before proper gun mounts were manufactured. (Roberts Collection)

Humber Armored Car Mark IV

The Humber armored car was prominent in the British Army with over 5,400 vehicles built. The Rootes Group manufactured the armored cars utilizing a design from the Guy Corporation. There were several versions of the Humber, but it was the Mark IV that was fitted with an American-made M5 or M6 37-mm gun. The larger 37-mm gun installed in this version of the Humber resulted in the removal of the radio operator to make room for the M5 or M6 gun mount. Approximately 2,000 vehicles of this type were produced. The vehicle was used extensively in the theaters where British forces were active.

Model of a Humber Mk IV armored car. (Roberts Collection)

T17 Deerhound armored car. (Wikimedia Commons)

T17 Deerhound Armored Car

The Ford T17 armored car and the Chevrolet T17E1 were medium armored cars submitted to the Ordnance Department in 1941. Each was equipped with the 37-mm M6 gun in an open turret. The Special Armored Vehicle Board decided that there were too many competing armored cars for standardization and canceled all programs except for the T22 which became the M8 armored car. The T17E1 became the Staghound and was given to the British. Approximately 25 T17s were completed as a temporary measure until the M8 production was ramped up. When M8s were in sufficient quantity, the 37-mm guns were removed from the T17s and the vehicles were used by the military police.

T13 Armored Car

In January 1941, the Trackless Tank Corporation proposed an eight-wheeled armored car to the U.S. Army that was tested at Aberdeen Proving Grounds. Thirteen of these vehicles were expected to be fitted with the 37-mm gun mounted in a turret that was similar to that on the M3 Lee medium tank, for further testing. Difficulties in production and scheduling persisted and because of the emergence of the M8 armored car, the T13 was considered unnecessary. The project was canceled in January 1943.

T13 armored car.
(Roberts Collection)

ARMORED FORCE BOARD 3357   P-264   FORT KNOX, KENTUCKY

Test of Armored Car T-13, No. 2. Profile. Note turret turned to rear.

T19E1 armored car. (Roberts Collection)

T19E1 Armored Car

The T19 armored car and its variants were designed by Chevrolet Corporation in 1942 to improve performance over the current 4x4 and 6x6 designs. Two prototypes were manufactured and tested but the overall performance was not significantly advantageous over existing designs. The vehicle had a 37-mm M6 gun mounted in a turret. In December 1942, the project was canceled.

T27 armored car. (Wikimedia Commons)

T27 Armored Car

The T27 armored car was developed by Studebaker Corporation in 1944 as a replacement for the M8 Greyhound armored car. The vehicle was tested at Aberdeen Proving Grounds and was found to have a degree of mobility superior to that of the M8 armored car. Further testing was performed at Fort Riley, Kansas, but after testing had been completed, it was deemed that there was no apparent need for additional armored cars and the project was canceled.

M38 Wolfhound armored car. (Wikimedia Commons)

M38 Wolfhound Armored Car

The M38 armored car was developed in 1944 as a replacement for the M8 armored car. The primary armament was the 37-mm M6 gun. The vehicle was powered by a Cadillac V8 engine through a Hydra-Matic automatic transmission. A few prototypes were produced and performed well during testing. With the end of the war in Europe, the U.S. Army canceled the project.

M38 Wolfhound armored car.
(Roberts Collection)

| 5 37-mm Gun With Non-Armored Vehicles, Aircraft, and Vessels in World War II

Production of the M3 37-mm gun on a carriage started in 1940. However, Army AT (antitank) units needed guns for training purposes. This problem was solved by Company A of the 93rd Infantry Battalion (AT) by taking parts from vehicles and making training guns using pipes as the gun barrels.

M3 37-mm Towed Mount M4

The training guns were modeled after the M3 and had many similar characteristics. The carriage axle assembly was a salvaged rear end and wheels from a 1930s civilian truck. Trails and other parts were fashioned from wood and bolted together. A Telescope M6 sight was used, if available, otherwise, a plain tube was used as a sight. Iron rod was forged into lunettes so that the practice gun could be towed by a Jeep or ¾-ton truck and give the drivers an opportunity to maneuver the gun using a vehicle. Training was effective, and units were ready to handle the actual guns when they arrived.

M3 37-mm gun with crew in firing positions. (Roberts Collection)

Under the April 1942 Army organization chart, each infantry battalion had an antitank platoon with four 37-mm towed guns hauled using Jeeps or ¾-ton trucks. Divisional artillery battalions had six 37-mm towed guns and combat engineering battalions had nine guns towed by M2 halftracks. Divisional headquarters company had four 37-mm towed guns (towed by ¾-ton trucks) and divisional maintenance company had two. By 1942, most of these towed guns were replaced by self-propelled guns as soon as they were made available. In 1942, the Airborne Division charts showed forty-four 37-mm antitank guns. Glider infantry regiments had a means of deploying the guns in gliders, while the parachute infantry units did not have antitank guns since deployment by parachute had not been developed. The 10th Mountain Division received antitank guns. Armored divisions had sixty-eight 37-mm antitank guns. By midwar, most of these guns were replaced with higher-caliber weapons such as the 57-mm antitank gun. The Marine Corps was using the 37-mm M1916 gun for training but had also received many of the M3 guns and the 37-mm GMC M6 motor carriage. Marine divisions were supplied with thirty-six 37-mm antitank guns. These proved effective against Japanese armor and bunkers.

Several other countries received the M3 37-mm antitank gun including the Chinese National Revolutionary Army, Bolivia, Canada, Chile, Colombia, Cuba, El Salvador, France, Paraguay, Britain, and Russia. The M3 was first used in the defense of the Philippines. During the Guadalcanal campaign, it knocked out several Japanese tanks and was effective against infantry (canister shot). Throughout the war in the Far East, the M3 37-mm antitank gun was effective against Japanese vehicles that were thinly armored. Japanese armored attacks were usually piecemeal with no large forces attacking defensive positions. Consequently, it was easy to pick off individual armored vehicles with the M3. The M3 was light enough to lift and carry over obstructions to the delight of the gun crews when they had to make a rapid change in placement. The M3 was used by the Marines until the end of the war.

37-mm gun M3 with the 70th Tank Battalion, 1940. (C. Roberts, *Armored Strike Force: The Photo History of the 70th Tank Battalion in World War II*)

Shown above are 37-mm gun crews from the 404th Engineer Company in New Guinea. (Roberts Collection)

37-mm fixed gun position of the 404th Engineer Company in New Guinea, circa 1943. (Roberts Collection)

M3 37-mm antitank gun on M4 carriage at Erie Proving Grounds at maximum gun tube elevation, November 21, 1941. Note the civilian truck tires, since military tires using non-directional tread are not necessary. (U.S. Army)

Experimental trailer hitch so that several gun carriages could be towed by one prime mover. This was never used in combat in World War II. (U.S. Army)

The experience with the M3 in North Africa proved that the gun was not sufficiently powerful to destroy the newer production Panzer III and IV tanks. The M3 was later replaced by the 57-mm antitank gun around early 1944. The M3 remained in use in Italy until late 1944, mainly because of the low priority of upgrading equipment in the Mediterranean Theater of Operations. By the time Allied units had crossed France, the M3 had been replaced by the 57-mm antitank gun.

37-mm Antiaircraft Automatic Gun M1A2

The 37-mm Antiaircraft Automatic Gun M1A2 incorporated the 37-mm M3 gun mounted on the M3A1 four-wheel carriage that could be towed at up to 50 mph on paved roads or anchored to the ground by raising the wheels. This gun was intended as an antiaircraft weapon but could also be used as an antitank gun. The combination M42 mount had the 37-mm M1A2 gun mounted in the center with two Browning .50-cal machine guns, one on each side.

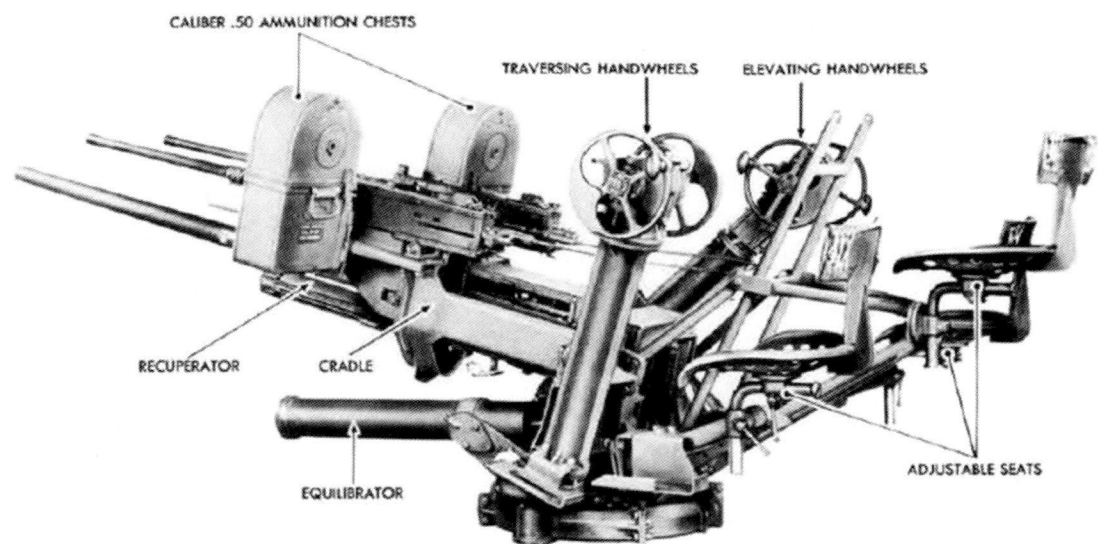

Combination M42 antiaircraft gun mount. (TM 9-235)

Combination M42 antiaircraft gun mounted in a halftrack, the CGMC T28E1, southern France, August 1944. (U.S. Army)

The 37-mm Automatic Gun M1A2 was recoil operated with an ammunition clip of 10 rounds. The cartridges were fed into the left side of the feed box and the shells ejected through an opening along the long axis of the gun. A new clip could be fed into the firing mechanism without interrupting the firing rate. The trunnion block is the housing of the gun which provides a mount for the gun, supports the gun tube, recuperator group, lock frame assembly, back plate group, driving spring assembly, and feed box group. The feed box is located approximately at the top of the trunnion block about halfway along its length. The feed box supports the 10-round ammunition clip. The gun tube is threaded into the tube extension. The tube can be replaced by the gun crew.

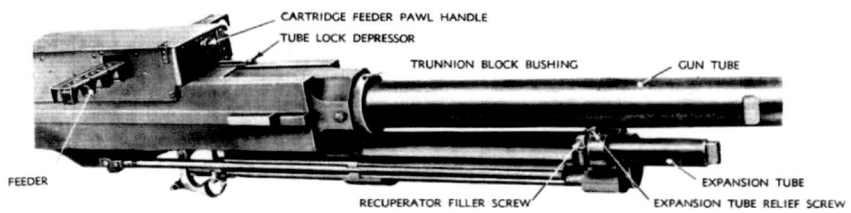

Automatic gun breech system (TM 9-235)

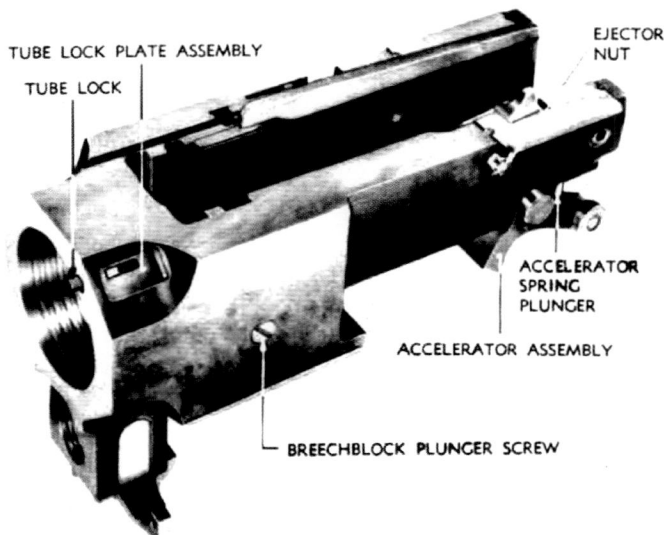

Gun tube extension. This section was added onto the breech end of the gun to accommodate the rapid-fire mechanism. (TM 9-235)

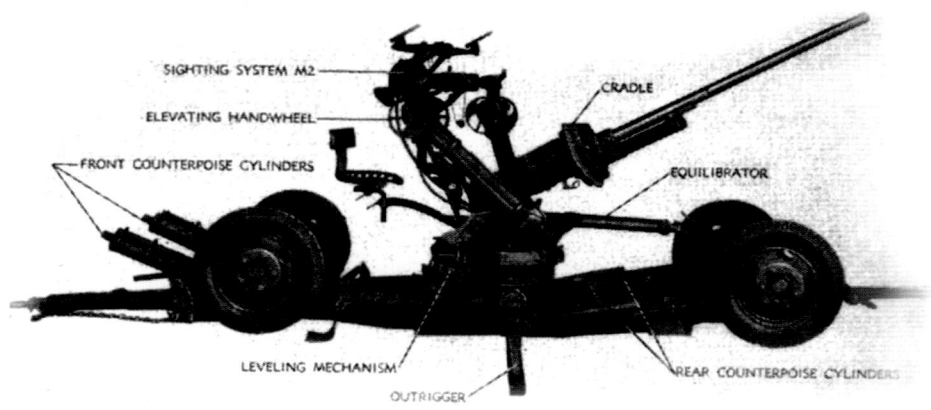

SIGHTING SYSTEM M2

ELEVATING HANDWHEEL

FRONT COUNTERPOISE CYLINDERS

CRADLE

EQUILIBRATOR

LEVELING MECHANISM

OUTRIGGER

REAR COUNTERPOISE CYLINDERS

37-mm AA gun M1A2 on carriage. (TM 9-235)

Specifications: 37-mm AA Gun M1A2

Bore, 37 mm .... 1.456 in

Weight of gun .... 365 lb

Length of gun .... 104 in

Weight of tube .... 119 lb

Length of tube .... 78 in

Rifling: uniform twist to right, one turn in .... 43.71 in

Number of grooves 12

Depth of groove .... 0.02 in

Width of grooves .... 0.2314 in

Width of lands .... 0.15 in

Number of turns in rifling in tube 1.56

Muzzle velocity .... 2,600 ft/sec

Rate of fire .... 120 rpm

Breechblock .... vertical sliding

Recoil mechanism .... hydro-spring

Max vertical range (HE cartridge) .... 6,200 yd

Max horizontal range (HE cartridge) .... 8,875 yd

Vertical range self-destroying HE cartridge .... 3,960 yd

Horizontal range self-destroying HE cartridge .... 4,070 yd

T2 37-mm gun motor carriage (gun mounted on a Bantam BRC 40 Jeep), 70th Tank Battalion, circa 1940. (C. Roberts, *Armored Strike Force: The Photo History of the 70th Tank Battalion in World War II*)

T2 37-mm Gun Motor Carriage

The T2 and T2E1 gun motor carriages were stopgap measures for immediately providing mobility for the M3 37-mm antitank gun. The T2 was a 37-mm antitank gun placed in a Bantam BRC-40 Jeep and could fire forward only. The gun was too heavy and the traverse was limited which led to the T2E1 version with rear fenders removed, a new mount, and 36° traverse. Further development of this vehicle was stymied by the inability of the chassis to support the weight of the gun and the gun mount.

The 37-mm antitank gun was the main antitank weapon prior to the war. At the beginning of World War II, it was proposed that the 37-mm antitank gun could be better utilized if it were mounted on a vehicle. It could be moved quickly and engage more targets easily when compared to the towed gun. This resulted in two development projects initiated by the War Department, the T2 37-mm gun motor carriage and the T8 37-mm gun motor carriage. After several unauthorized experiments with the 37-mm gun mounted on Jeeps, the Ordnance Department sanctioned development of the T2 37-mm gun motor carriage using the Bantam BRC-40 Jeep as the transport vehicle

Drivers testing the T2 felt it to be a little top heavy, probably from the weight of the gun. The Bantam BRC had a gross weight of 2,080 lb, but with the 37-mm gun, ammunition, and crew, the

T2E1 37-mm gun motor carriage with gun mounted on a Bantam BRC 40 Jeep, 70th Tank Battalion, circa 1940. (C. Roberts, *Armored Strike Force: The Photo History of the 70th Tank Battalion in World War II*)

T2 37-mm gun motor carriage. (Roberts Collection)

gross weight was 3,510 lb, way beyond the design specifications of the Bantam BRC. The T2E1 with parts of the body removed and two crew resulted in a gross weight of 3,250 lb. The pedestal mount on the T2E1 allowed firing of the 37-mm gun to the side, but when this happened, the tires on the muzzle side would lift off the ground. The 37-mm antitank gun is a high-velocity gun (2,900 ft/sec) and packs a significant recoil. This caused the T2 to jump or rock excessively making it difficult for the gunner to reacquire the target. The overloading of the vehicle and the instability of the vehicle while firing were factors that resulted in the Armored Force Board canceling the project. All guns were removed and the Jeeps returned to normal service. All guns were sent back to ordnance to be issued to other applications.

T8 37-mm Gun Motor Carriage

The T8 gun motor carriage was the first vehicle designed specifically to carry the 37-mm antitank gun. The vehicle was manufactured by Ford Motor Company using components from the standard 1½-ton truck (Burma Jeep). The standard Ford six-cylinder engine was in the middle of the vehicle to the right side of the driver. This allowed the gun to be mounted in the front of the vehicle. The gross vehicle weight was 4,520 lb with 100 rounds of ammunition and a three-man crew. The original traverse of 30° right or left was retained for the 37-mm antitank gun along with the –10° to +15° elevation. The vehicle was stable when using the 37-mm gun, allowing the gunner to reacquire the target readily. Because of the large tires, cross-country performance of the T8 was reasonably good. Steering and cooling problems were discovered during testing but later corrected. With no windshield, the crew had to wear special clothing to keep warm. The driver would remain in position while the gunners either sat in the seats on the right or rear of the vehicle when not engaging targets or kneeling next to the gun when firing. One gunner would load while the other would aim the gun with the help of the driver who would align the vehicle in the general direction of the target. Although this was intended to be an antitank vehicle, it offered very little protection for the crew. Development of this vehicle ceased in April 1942 because of the testing on the T21 (M6 GMC) which used a production vehicle, the WC 52, for transport.

M6 37-mm gun motor carriage in Sicily with the 70th Tank Battalion. (C. Roberts, *Armored Strike Force: The Photo History of the 70th Tank Battalion in World War II*)

M6 37-mm Gun Motor Carriage (T21)

In mid-1941, the Chrysler Corporation proposed a transport vehicle for the 37-mm antitank gun which was essentially the WC 52 weapons carrier already in production. The Ordnance Department liked the concept, designating the vehicle the T21 motor carriage. A pedestal was mounted in the box of the vehicle approximately over the rear axle. Reinforcement of the floor and frame was required in this area. The manually loaded M3 37-mm was chosen over the M6 since the automatic feature of the M6 did not offer a substantial improvement in loading times. The Ordnance Department felt that the T8 was better than the T21 but due to the availability of the WC 52 vehicles, the T21 could be developed as a bridge to the M8 armored car.

In early 1942, the T21 was designated 37-mm Gun Motor Carriage M6. During development, several modifications were experimented with, including a recommendation of a gas deflector, which later proved a problem with canister ammunition. The spare tire was eliminated. In September 1943, the 37-mm Gun Motor Carriage M6 was classified as a limited standard and several were converted back to the WC 52 weapons carrier. Some served in North Africa and Sicily, but it was never used in combat in Europe because of the inability of the 37-mm gun to penetrate the new German tanks. Several M6 GMCs were given to the Marines who used it effectively against the lightly armored Japanese tanks and bunkers.

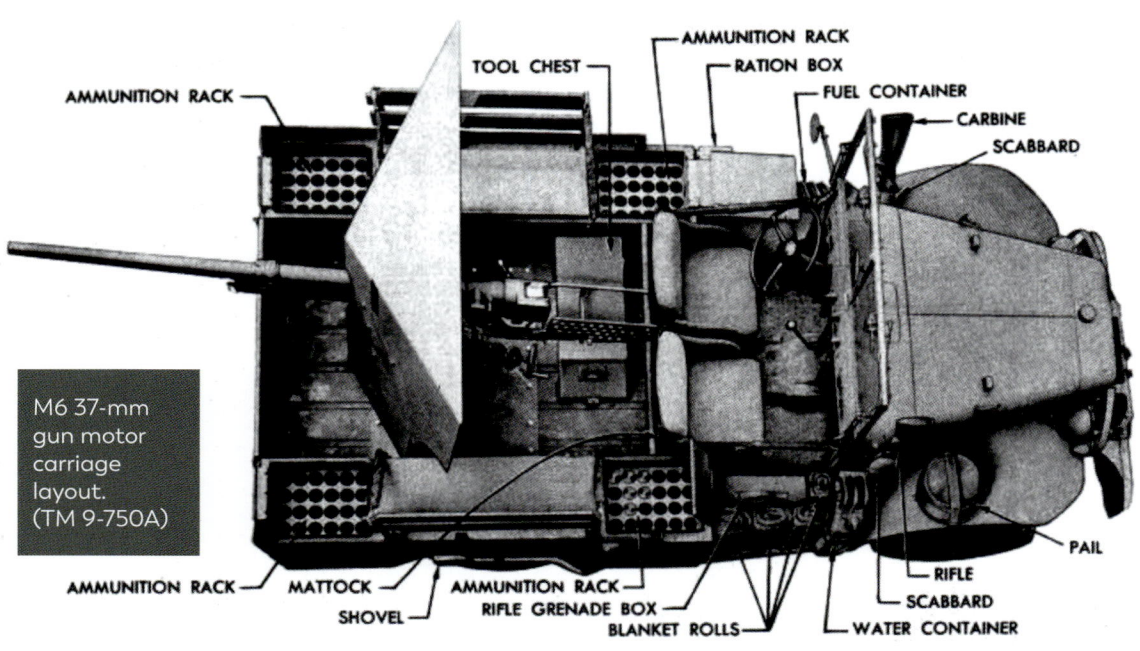

AMMUNITION RACK — TOOL CHEST — AMMUNITION RACK — RATION BOX — FUEL CONTAINER — CARBINE — SCABBARD

AMMUNITION RACK — MATTOCK — SHOVEL — AMMUNITION RACK — RIFLE GRENADE BOX — BLANKET ROLLS — WATER CONTAINER — SCABBARD — RIFLE — PAIL

M6 37-mm gun motor carriage layout. (TM 9-750A)

View of the breech on a restored M6 37-mm gun motor carriage. (Roberts Collection)

Left front view of a restored M6 37-mm gun motor carriage. (Roberts Collection)

Left rear view of a restored M6 37-mm gun motor carriage. (Roberts Collection)

In this photo, the 2nd Armored Division removed the 37-mm gun and mount from the M6 37-mm gun motor carriage and placed it in a halftrack. (U.S. Army)

37-mm Gun Motor Carriage T13 and T14

The T2 transport vehicle, the Bantam BRC 40, was considered overloaded, so an idea was put forward by Willys Overland to extend the Willys Jeep chassis to a 6x6 vehicle with higher load-carrying capability. The initial design, designated T13, had the 37-mm gun facing forward. The using military preferred the rear-facing gun T14 so the T13 was never constructed. In January 1942, a prototype (T14) was delivered to Aberdeen Proving Grounds. The 37-mm gun was mounted between the two rear axles. The vehicle was well received by the military in that the vehicle had reasonable mobility, small size, low silhouette, and easier maintenance since it used several parts from the Willys Jeep. At that time, the M6 37-mm gun motor carriage was being produced, and the Ordnance Department considered it unnecessary to produce another small 37-mm gun motor carriage, so the project was canceled in May 1943.

T14 37-mm gun motor carriage left side view. (Wikimedia Commons)

T14 37-mm gun motor carriage left front view. (Wikimedia Commons)

37-mm Gun Motor Carriage T33

In the search for a feasible 37-mm gun motor carriage, another concept using the Ford GTB chassis and parts was proposed which resulted in the 37-mm Gun Motor Carriage T33. It was similar to the T8 but had the gun mounted in the back and essentially looked like the Ford GTB already in production. Testing showed the vehicle to be satisfactory but duplicative of the T8 and other similar designs. Since the M6 GMC was already in production, no further testing was performed on the T33 and the project was canceled in April 1942.

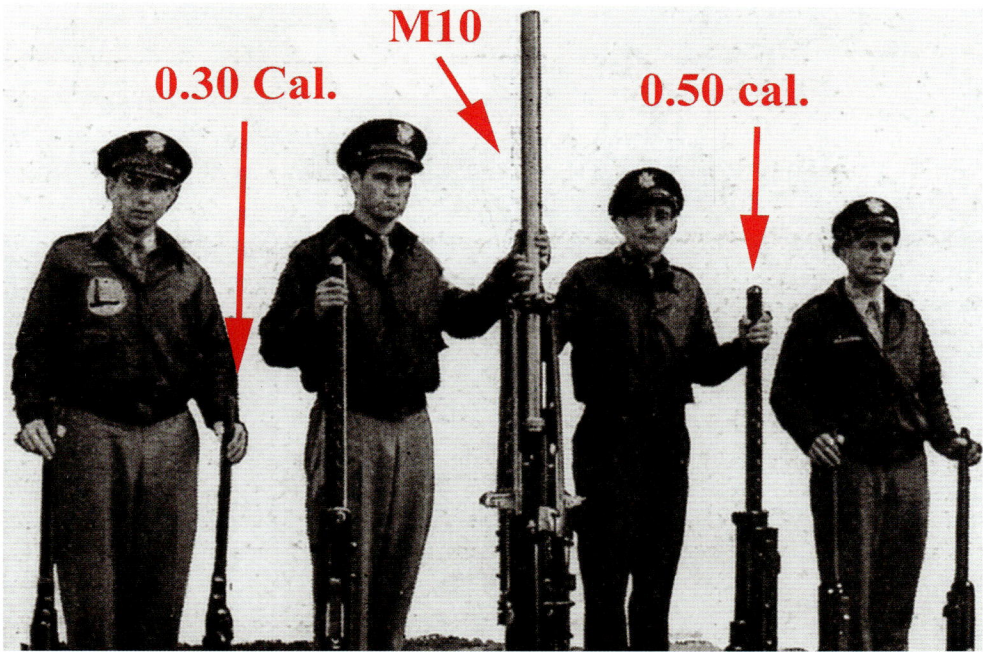

Comparison of automatic gun M10 with .50-cal. and .30-cal. machine guns. (U.S. Army)

37-mm Gun AN-M4 and M10 in Aircraft

The P-39 Airacobra and P-63 Kingcobra were World War II fighter planes with 37-mm guns mounted on the center line such that the barrel protruded through the propeller spinner (propeller hub). The 37-mm gun was not designed to fire synchronized with the propeller blades.

There were two versions of the gun with the difference being how the ammunition was supplied to the breech. The M10 gun used a disintegrating link belt, very much like a typical machine gun. The AN-M4 gun utilized a 30-round endless belt whose links did not disintegrate or fall away. Both guns were automatic firing, utilizing a solenoid connected to the cockpit via wiring. It was advantageous to have a gun mounted in the propeller spinner. It was easier to aim since the pilot could fire at varying ranges without the added chore of estimating the range required for guns outboard of the fuselage to converge on the target. The low velocity, around 2,000 ft/sec, resulted in projectile drop at long ranges, making it necessary to accommodate the drop by aiming high.

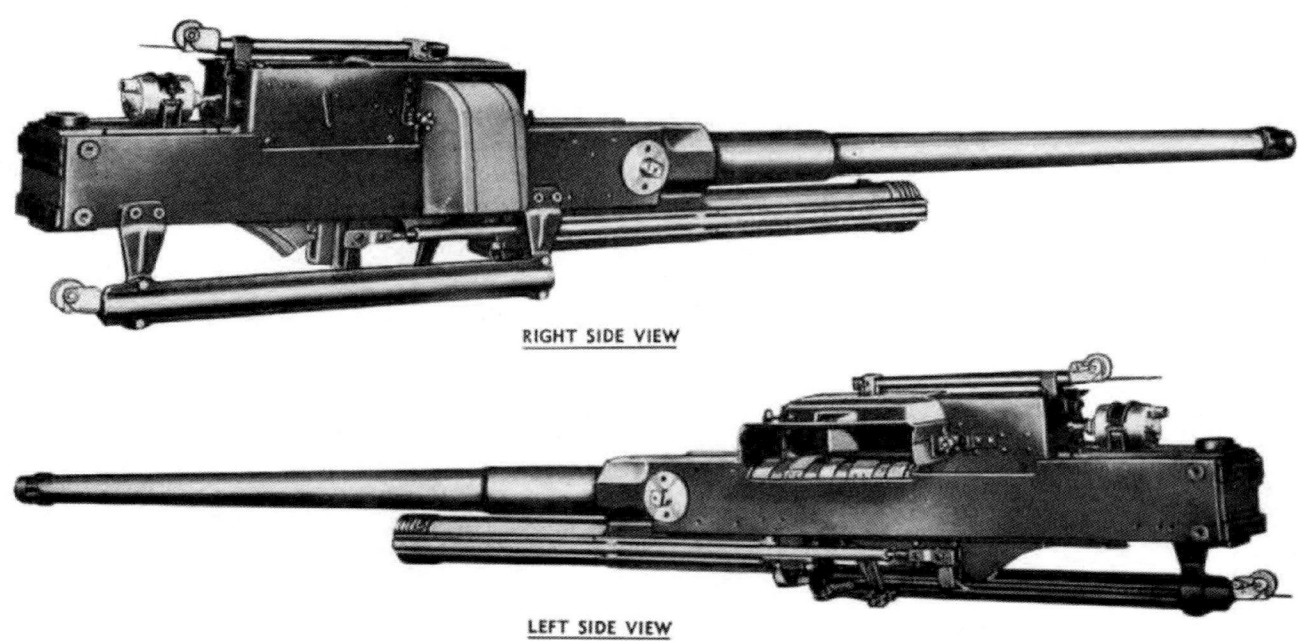

M10 gun right and left view. (TM 9-240)

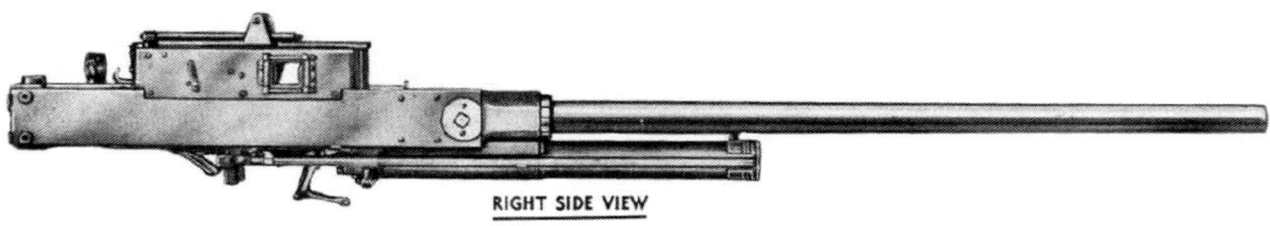

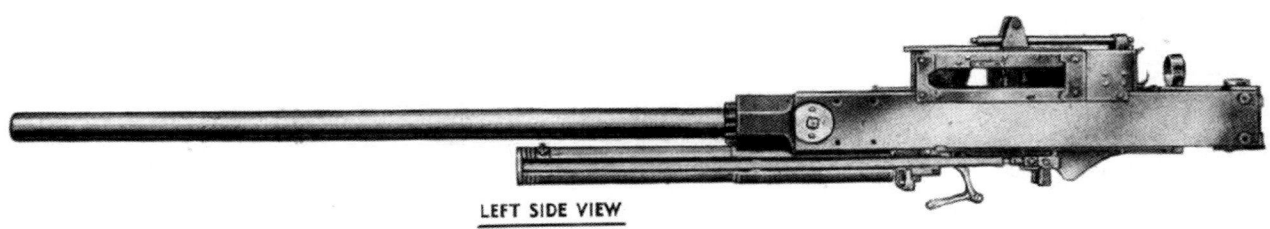

37-mm AN-M4 auto cannon by Browning. (TM 9-240)

AN-M4 automatic gun with endless-belt ammunition rack. (Wikimedia Commons)

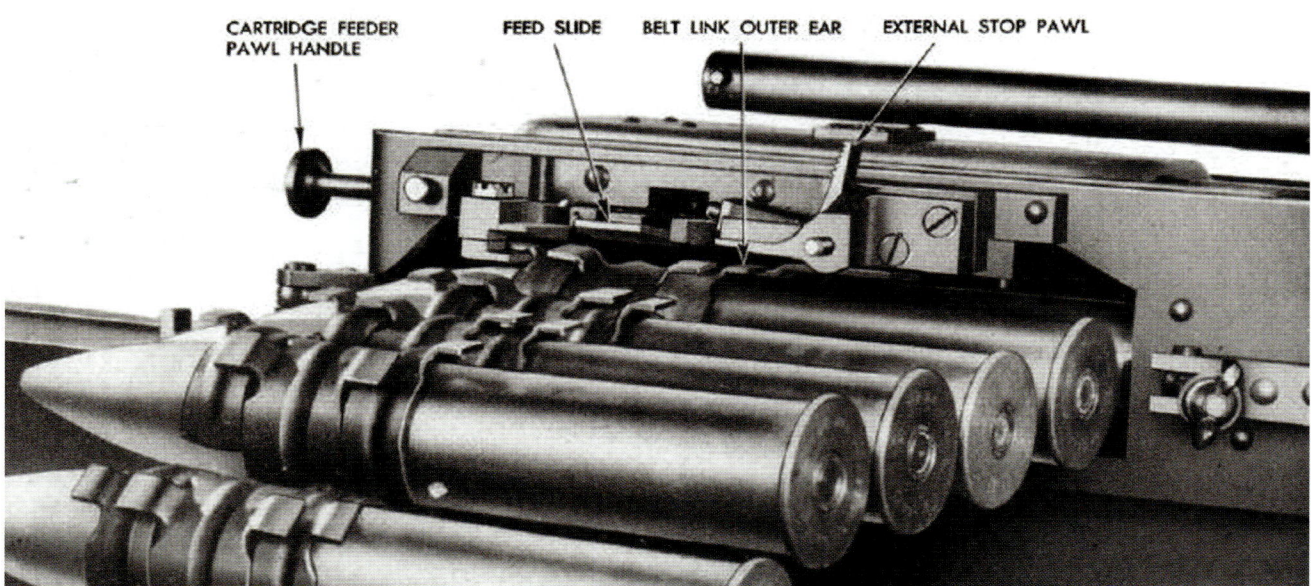

CARTRIDGE FEEDER PAWL HANDLE   FEED SLIDE   BELT LINK OUTER EAR   EXTERNAL STOP PAWL

M10 automatic 37-mm gun with disintegrating-link belt feed. (U.S. Army)

Patrol Torpedo (PT) Boats

In mid-1942, the PT boat arm of the U.S. Navy started installing AN-M4 37-mm guns using the endless-belt ammunition rack on the foredeck of torpedo boats. The intended target was Japanese landing barges delivering supplies and troops to various islands. These barges were not heavily armed and were vulnerable to the 37-mm projectiles. Initially, someone got the idea to cannibalize crashed P-39s from Henderson Field on Guadalcanal and mount their guns on the foredeck with a custom-made mount in the field. This setup was so successful that special mounts were designed and guns supplied as standard equipment. Field-expedient handgrips were manufactured in several styles and tracer ammunition was used to direct the shot at the enemy barges. Various gunsights were tried, but the tracer aiming worked the best. In 1944, the M9 37-mm gun (see table on page 51) was installed as standard equipment.

In Profile:
PT Boat & Browning AN-M4 37-mm Auto Cannon

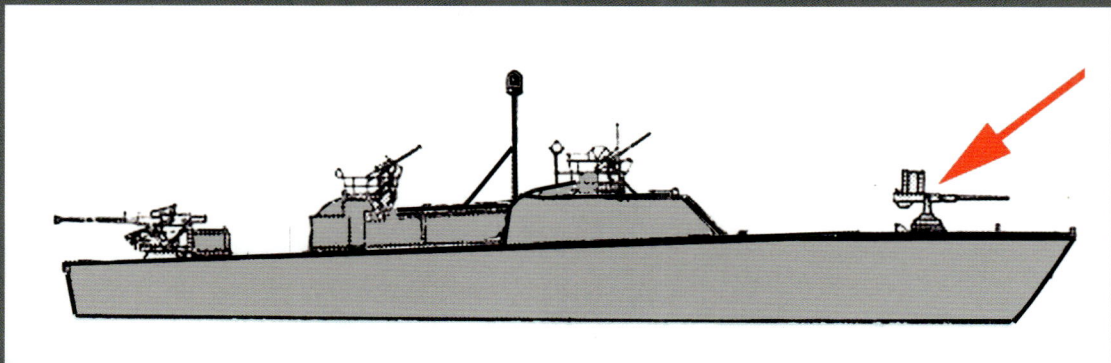

Patrol Torpedo (PT) boat with 37-mm auto cannon on the foredeck (red arrow). (Roberts Collection)

Browning M4 37-mm auto cannon with 30-round endless-belt ammunition rack on the foredeck of a PT boat. (Roberts Collection)

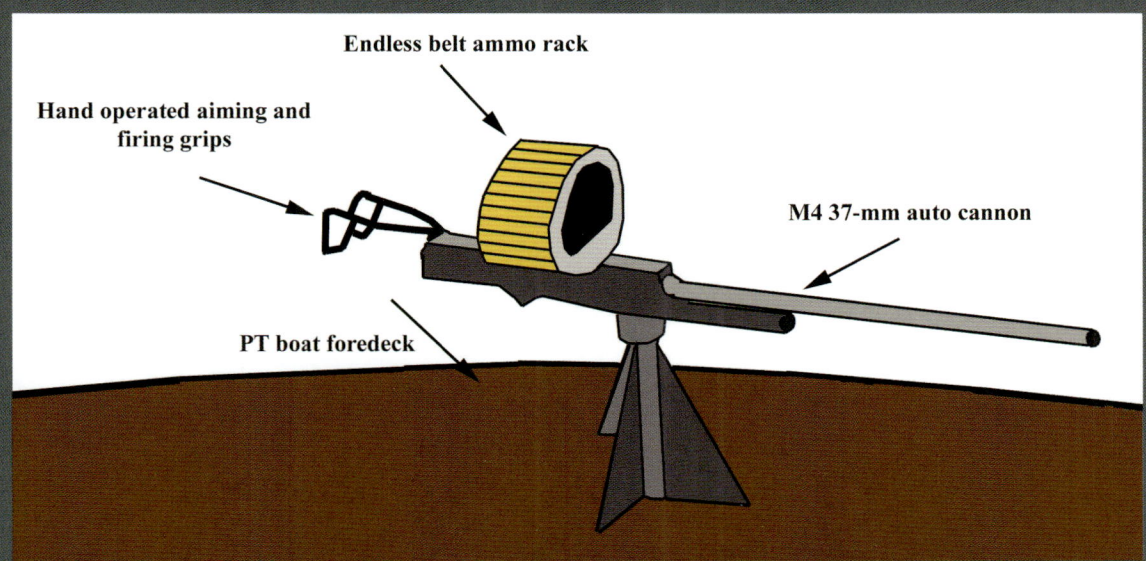

PT boat showing 37-mm M4 automatic gun on foredeck with 30-round endless-belt ammunition rack. (U.S. Navy)

Conclusion

The American 37-mm gun operated as designed, delivering 1.5- to 2-lb projectiles at velocities from 2,000 to 3,000 ft/sec. It was used on several armored vehicles, soft-skinned vehicles, aircraft, and PT boats. This attests to its utility as an effective weapon, despite the fact it could not defeat every target in the inventory of the enemy. It was easy to maneuver the weapon and fire it at a target. The American 37-mm gun has its place in the history of World War II as one of the most versatile weapons produced.

The Japanese Daihatsu-type landing craft was vulnerable to the 37-mm gun mounted on PT boats and as such was a typical target. (C. Roberts, *The Boat That Won the War*)

Appendix I

Design Drawings of M3 37-mm Gun and M4 Carriage Components

Produced by the Ordnance Office, U.S. War Department

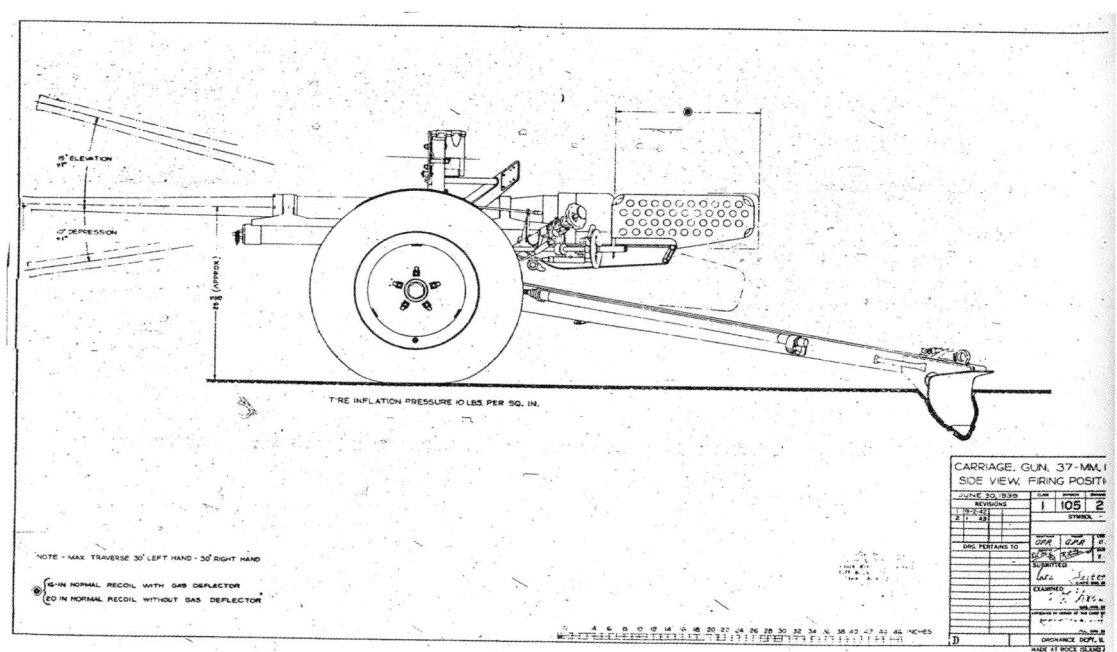

Side view of Carriage, Gun 37-mm.

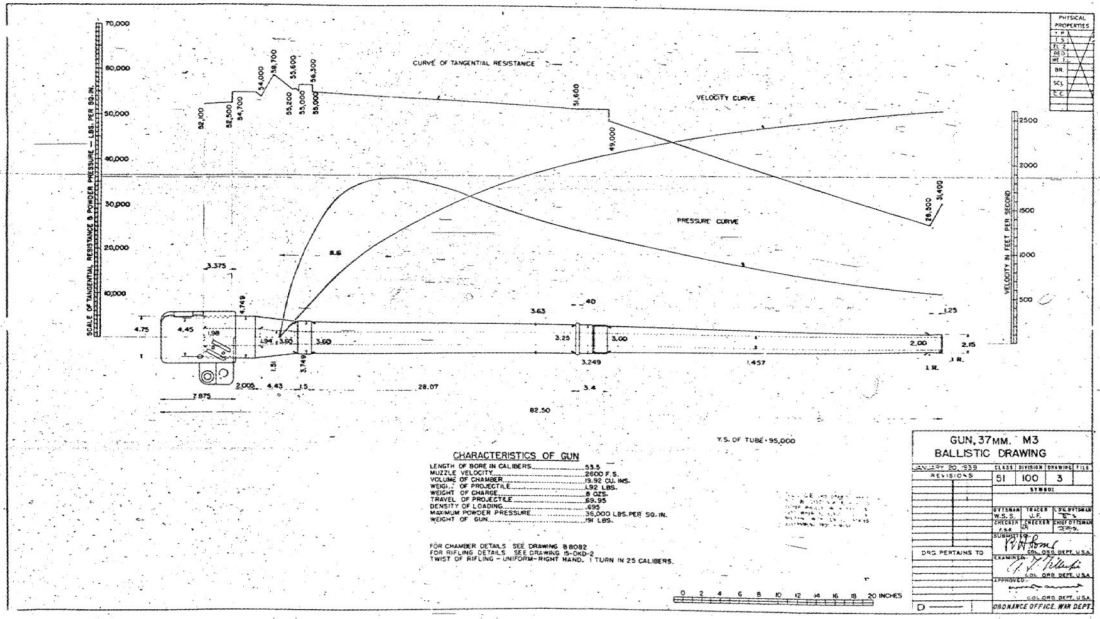

Gun, 37-mm M3 Ballistic Drawing.

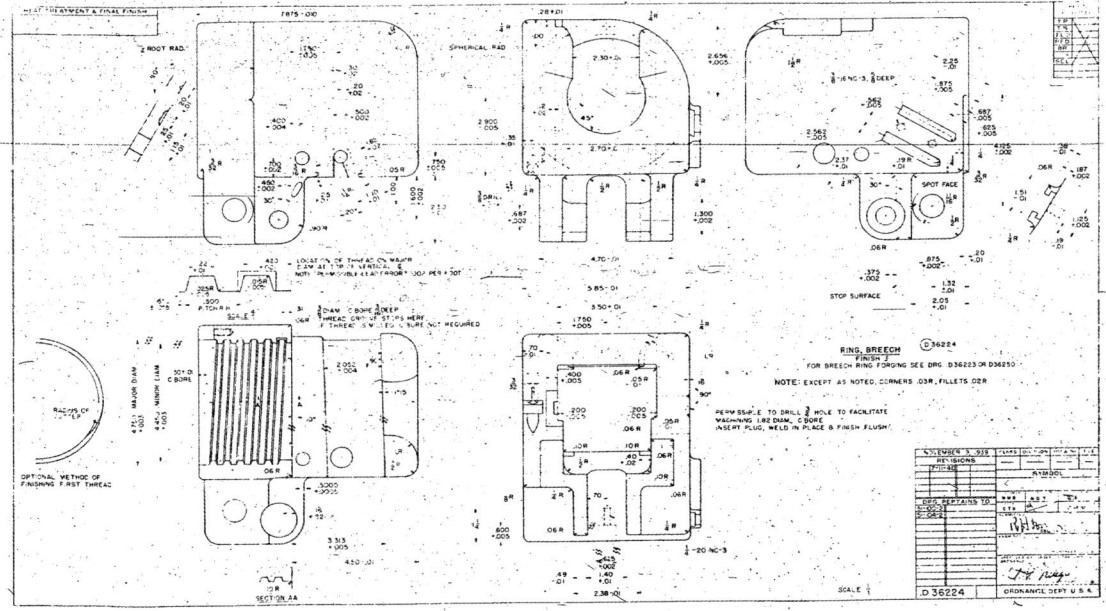

Breech Ring Forging, 37-mm M3.

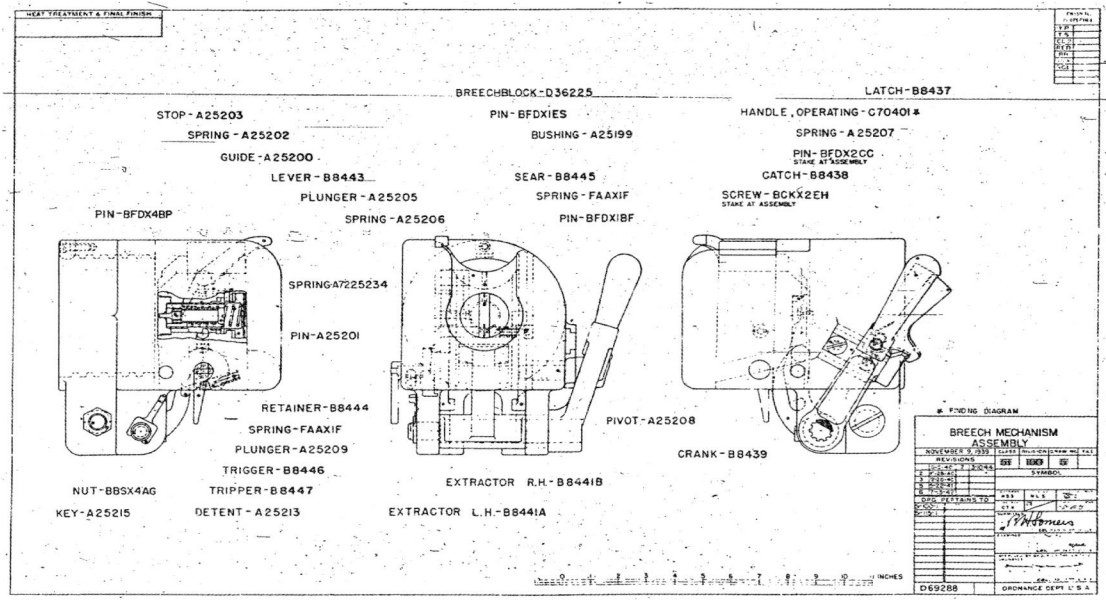

Breech Mechanism Assembly, 37-mm M3.

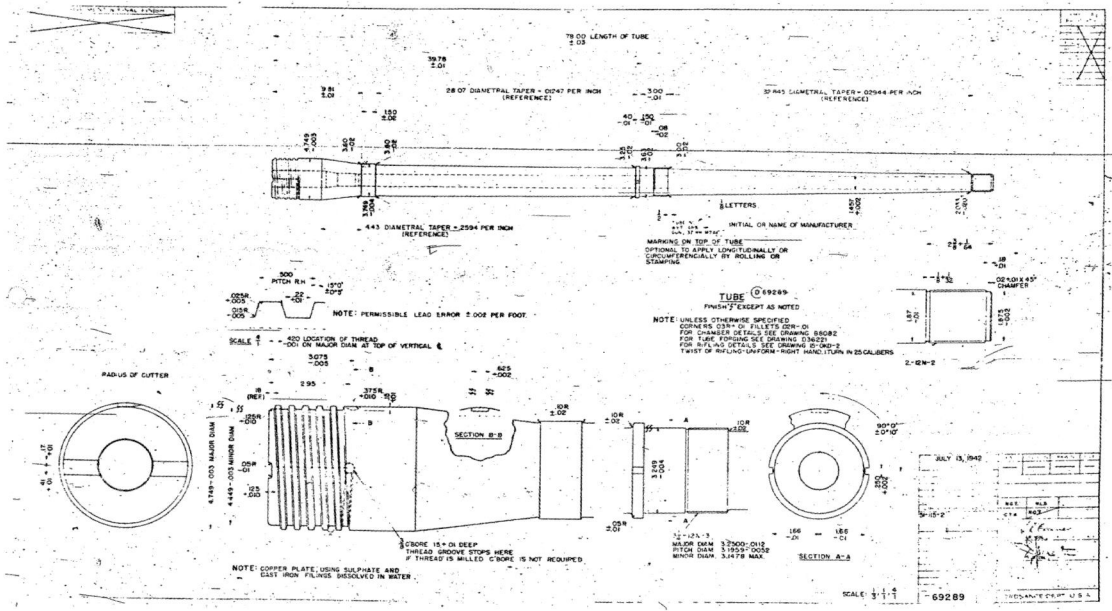

Gun Tube and Chamber Detail, July 13, 1942.

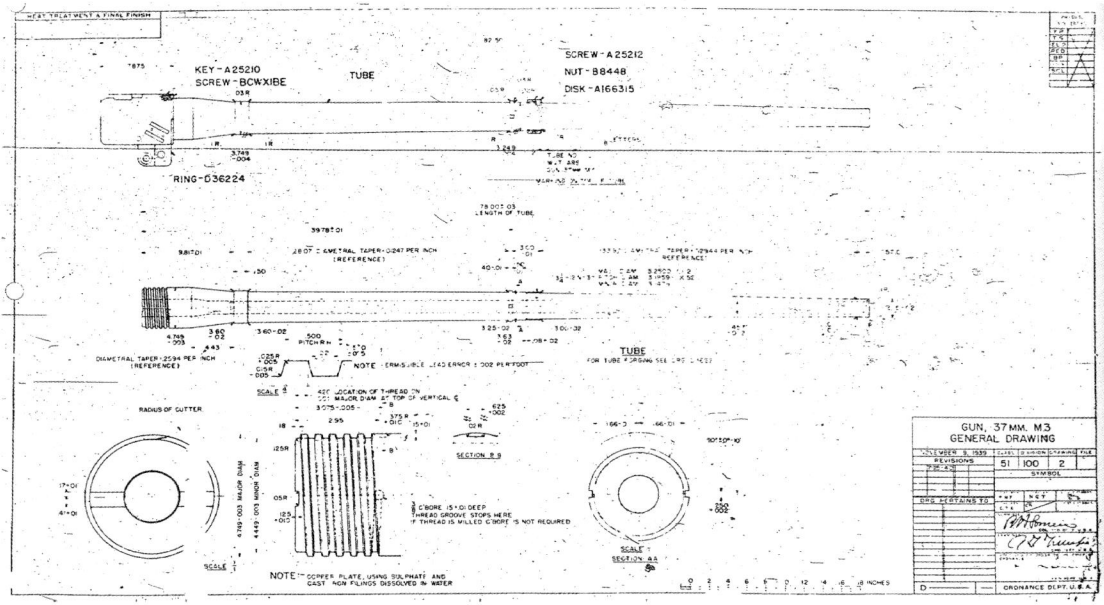

Gun, 37-mm M3 General Drawing.

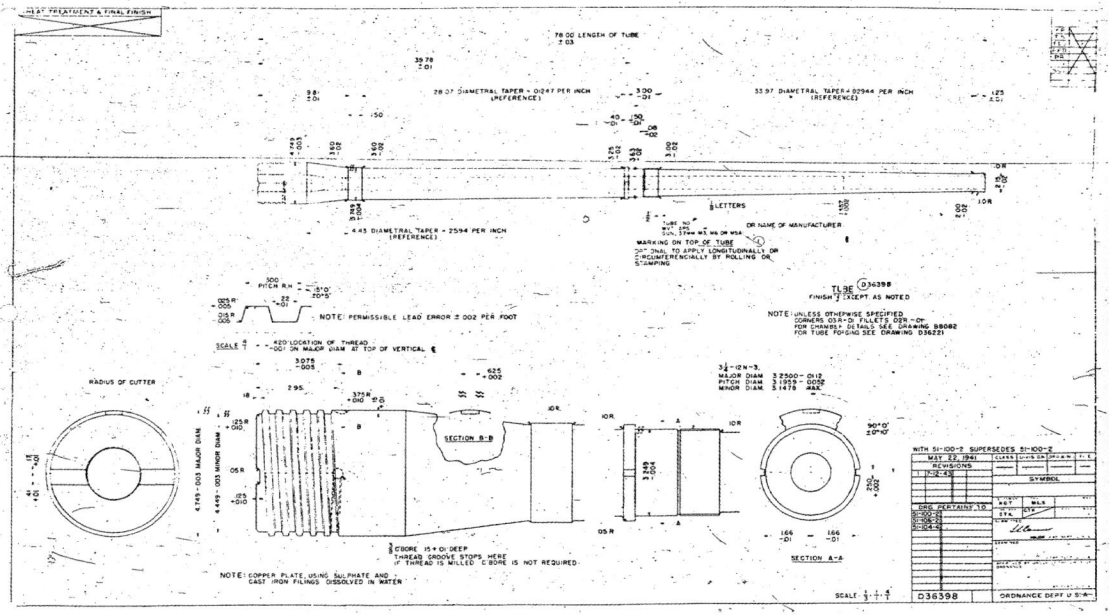

Gun Tube and Chamber Detail, May 22, 1941.

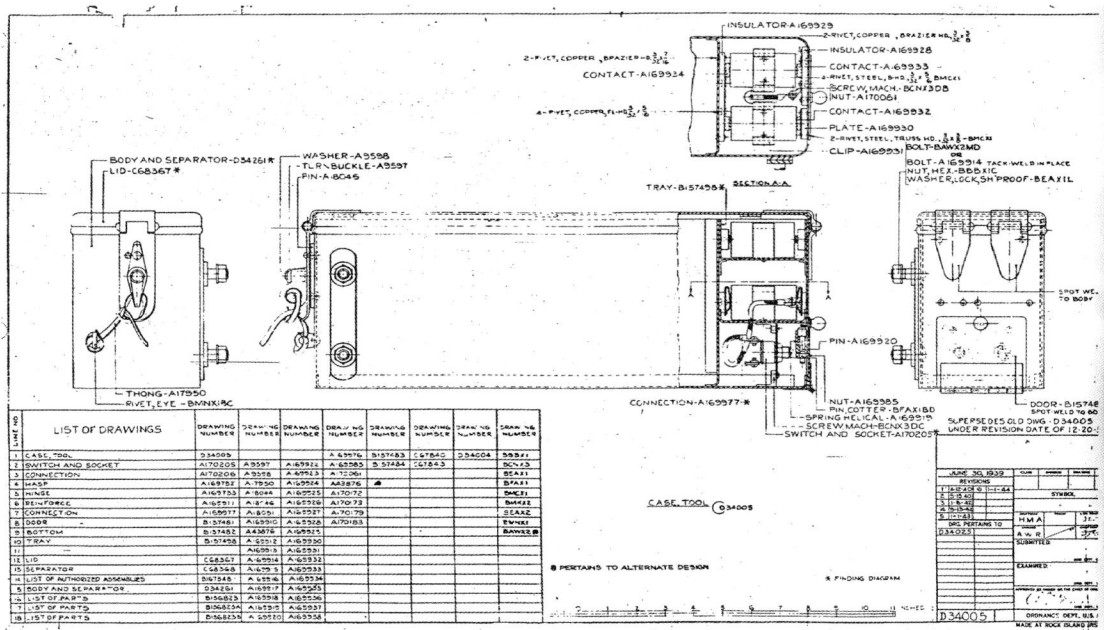

Tool Case.

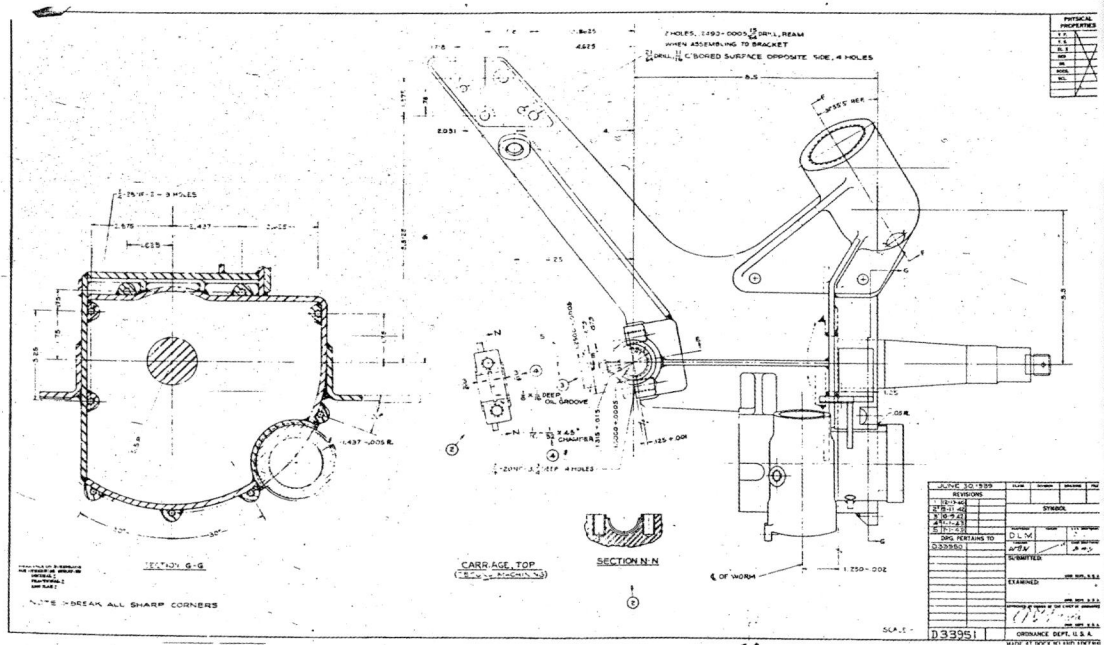

Carriage, Top M4.

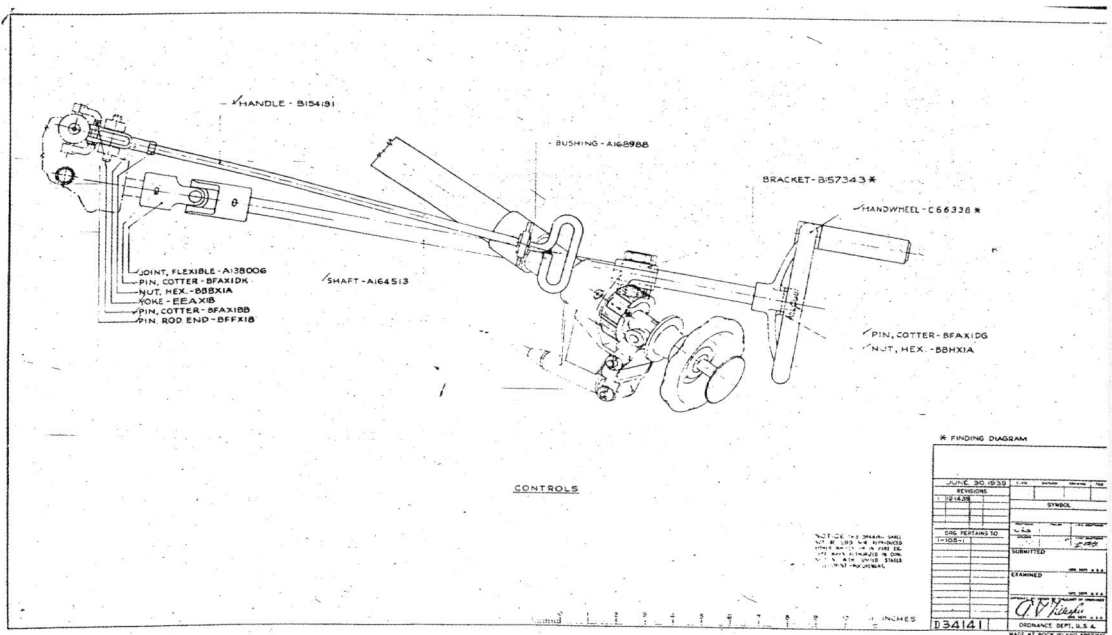

Controls.

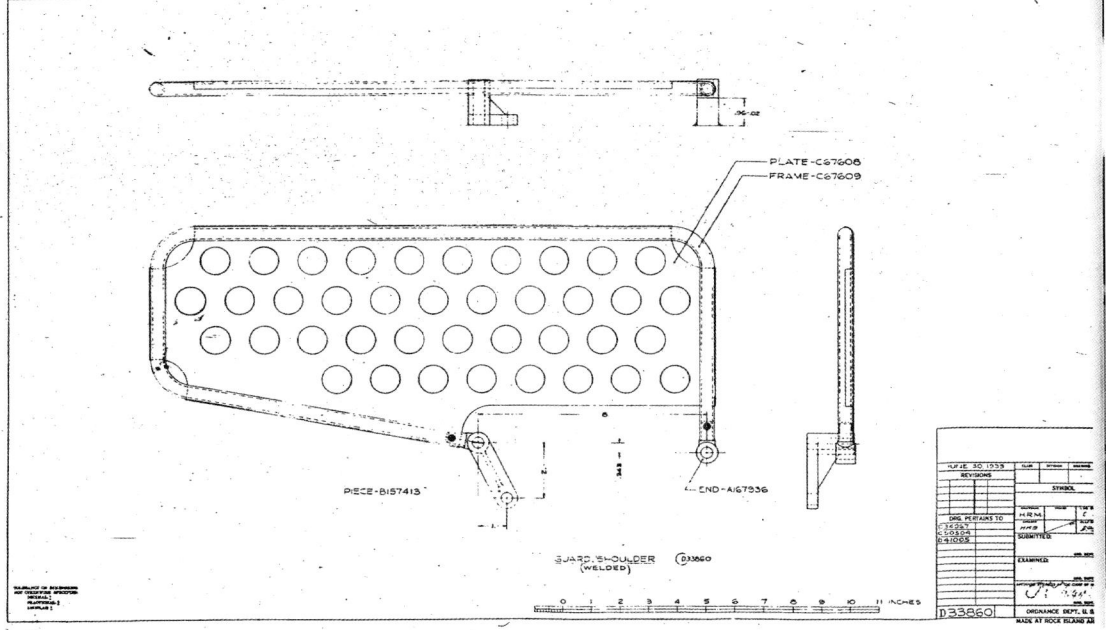

Shoulder Guard.

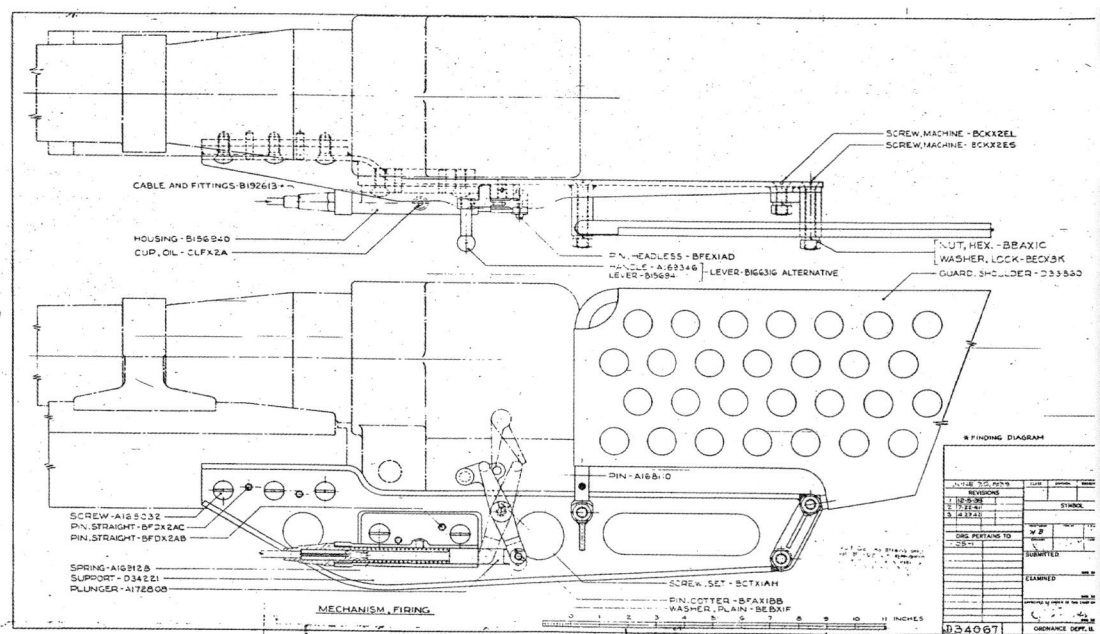

Firing Mechanism, 37-mm Gun M3.

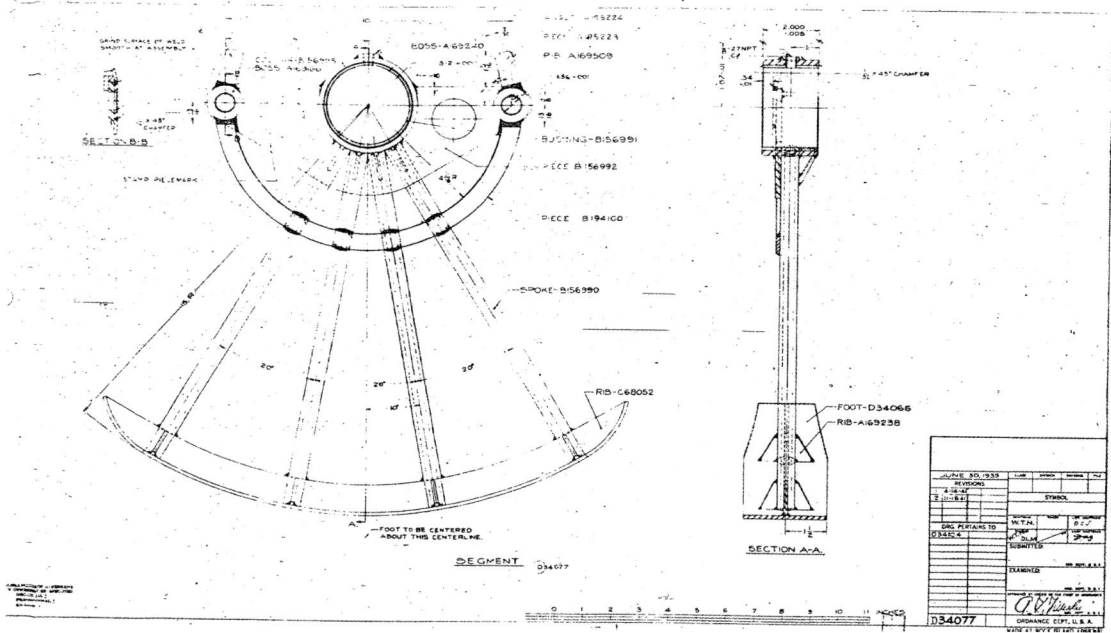

Wheel Support Segment, M4 Carriage.

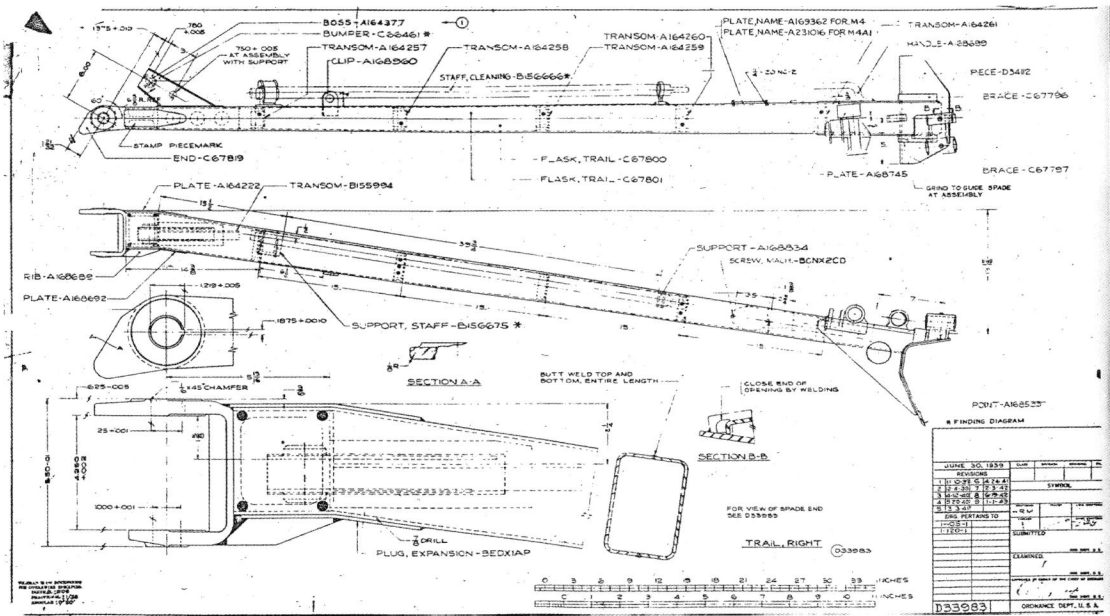

Right Trail Arm, M4 Carriage.

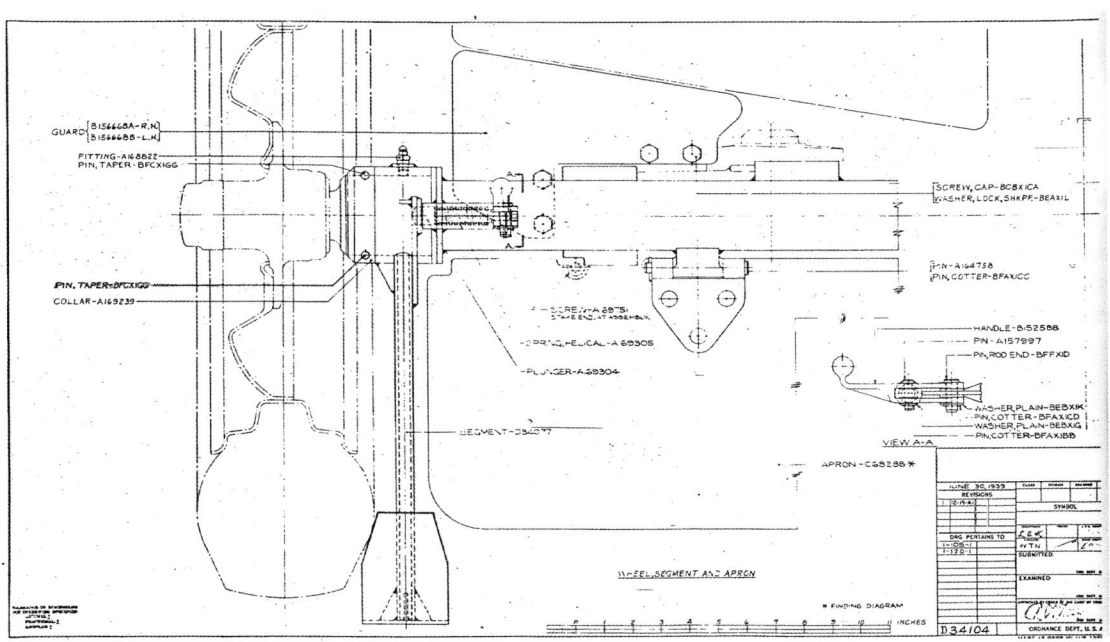

Wheel Segment, Apron, Wheel and Axle, M4 Carriage.

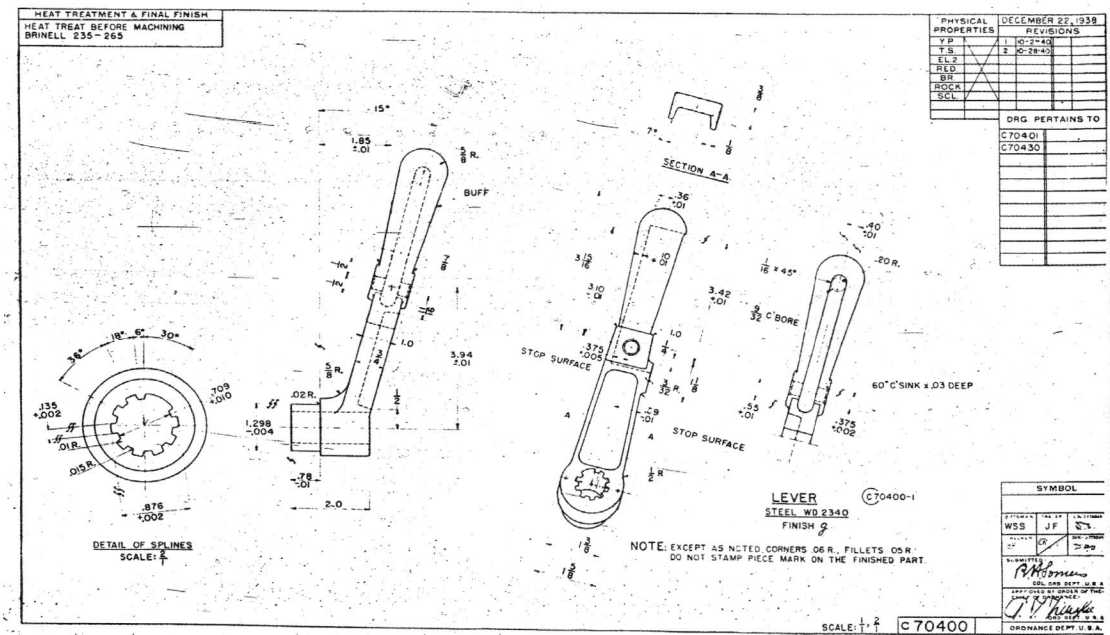

Spent shell ejection lever attached to the breech dropping block.

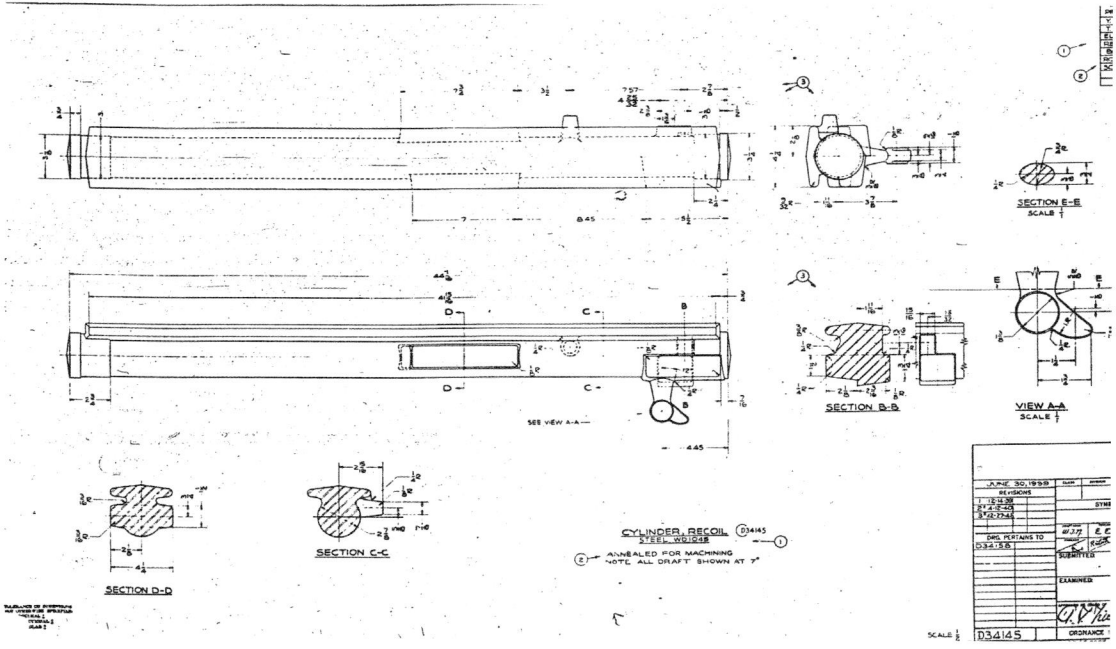

Recoil cylinder.

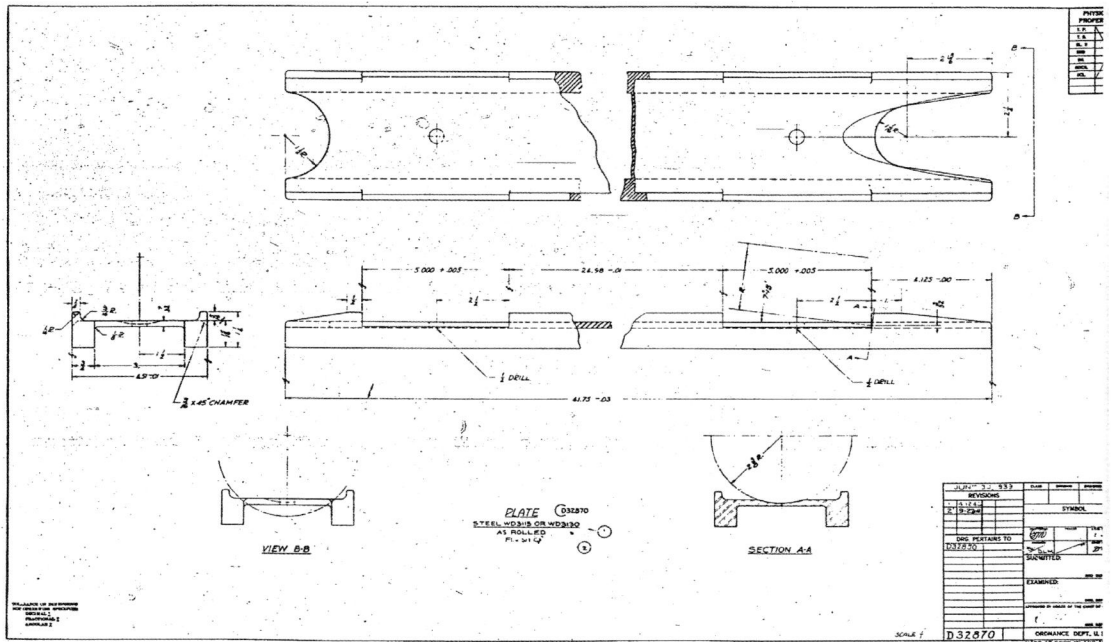

Steel gun support slider plate.

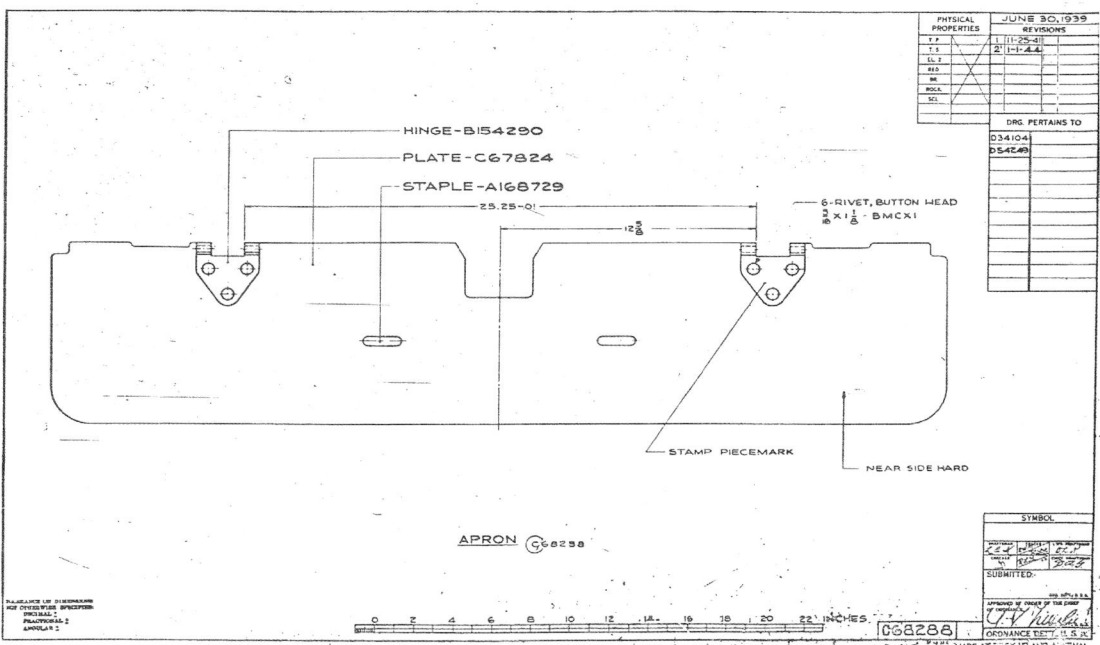

Lower armor apron plate attached to main armor plate. This is folded horizontal for traveling.

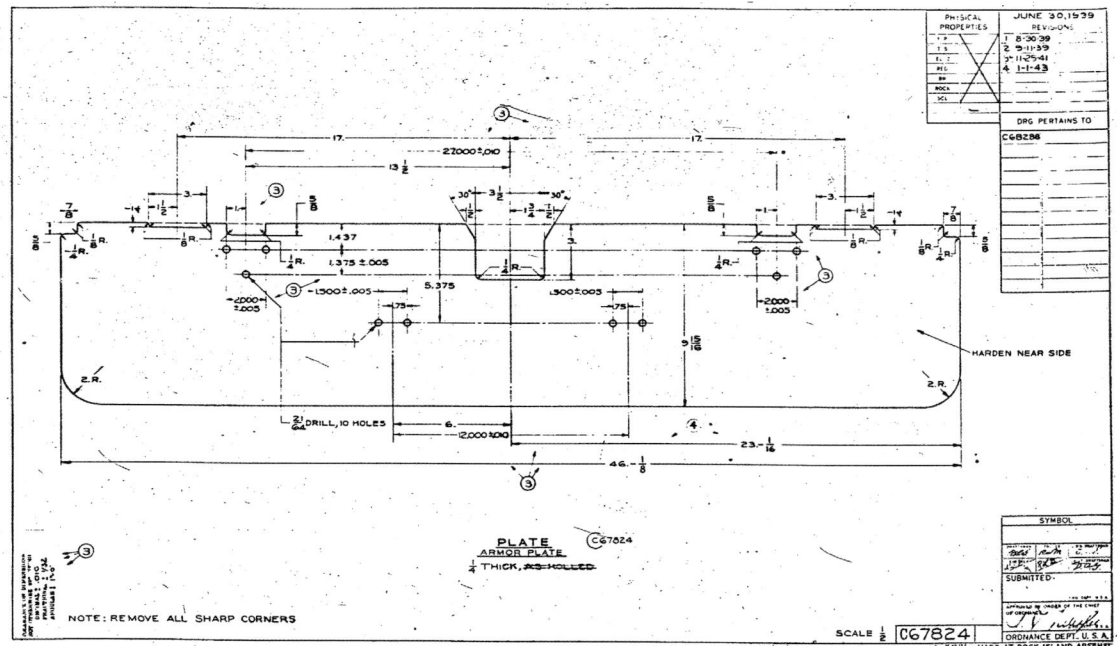

Detailed drawing of lower armor apron plate attached to main armor plate.

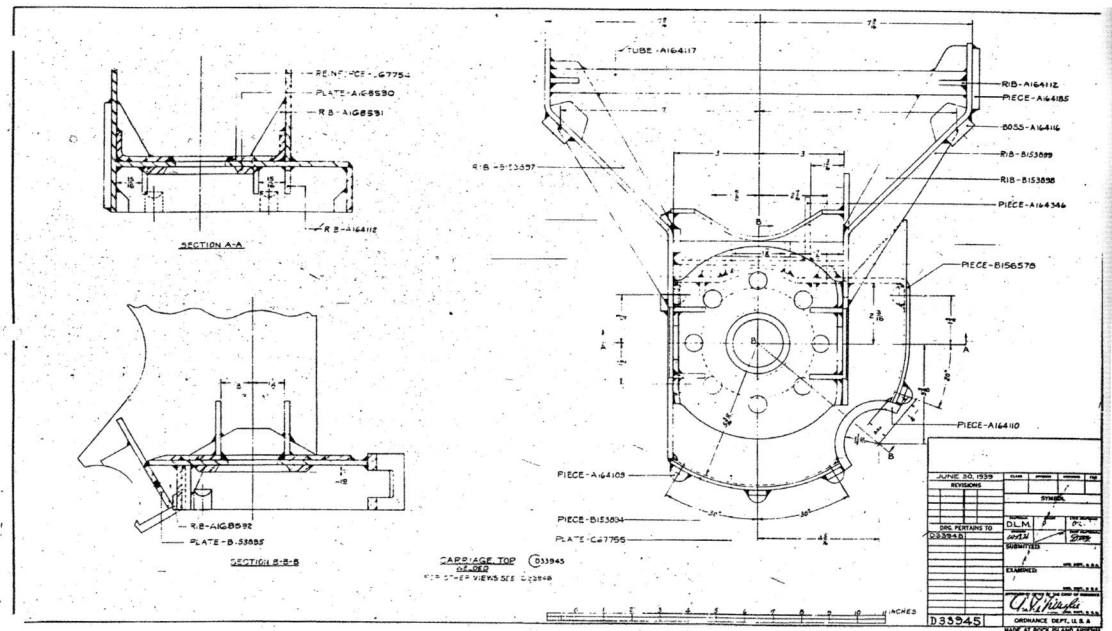

Top of gun carriage assembly.

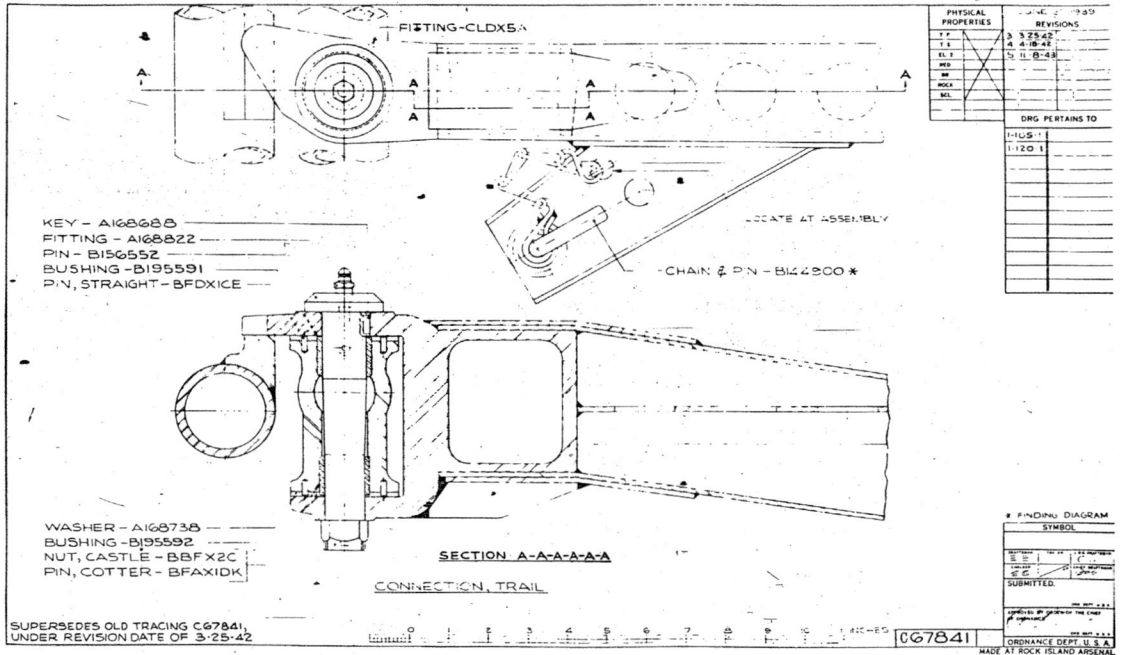

Gun carriage trail connector.

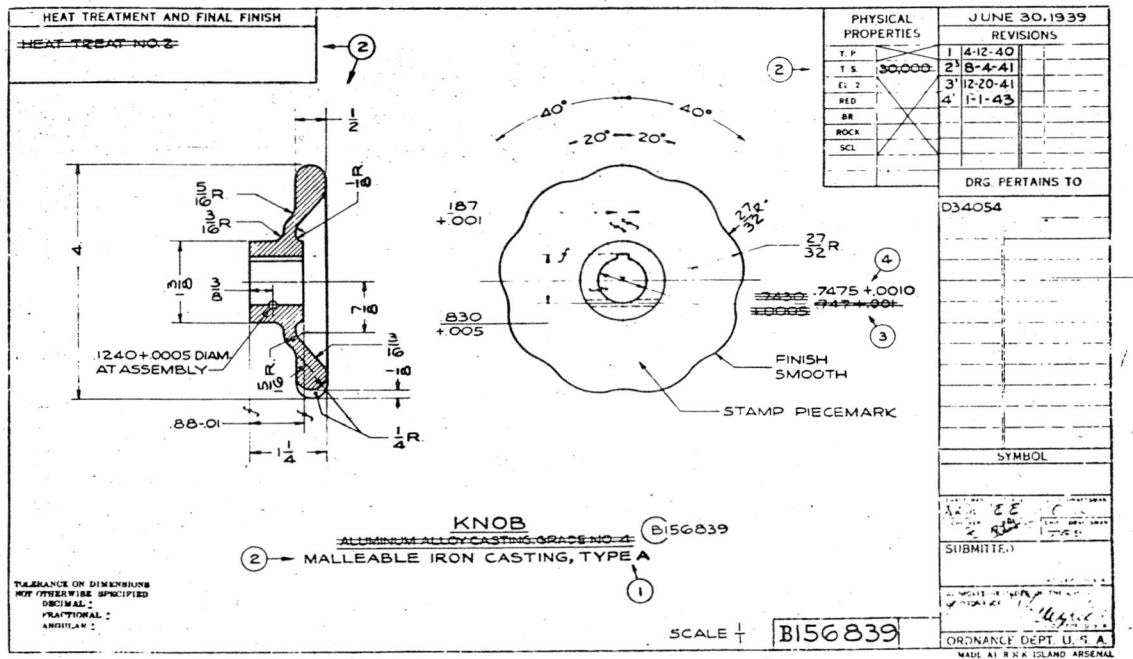

Cast iron gun elevation knob.

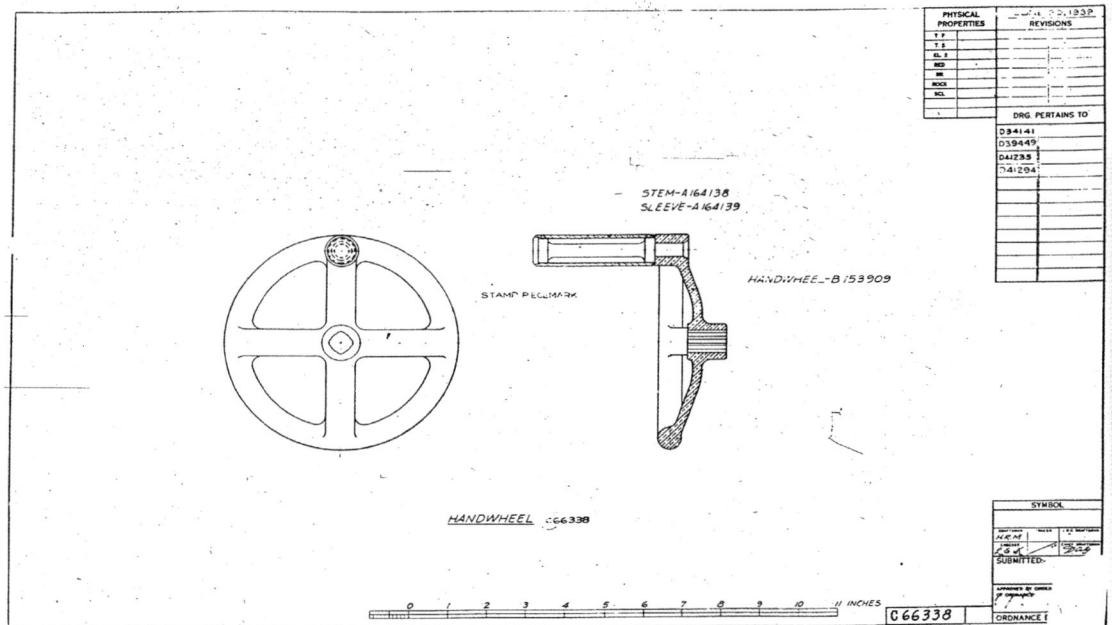

Handwheel for gun elevation or traversing depending on the gun model.

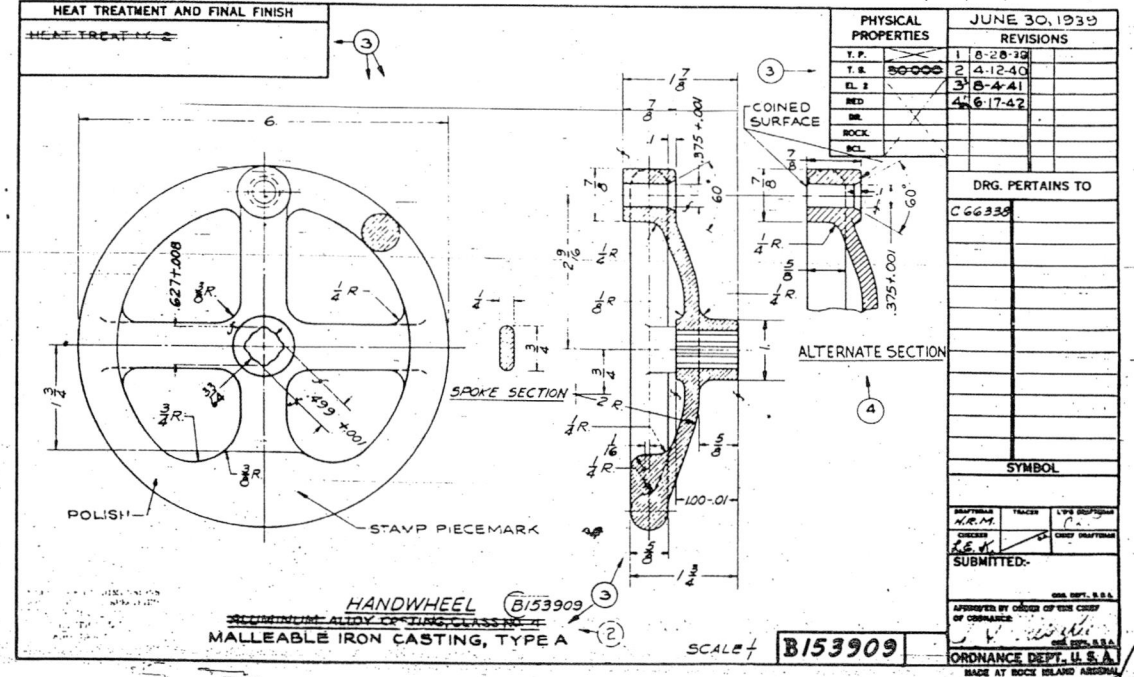

Detailed drawing of hand wheel shown above.

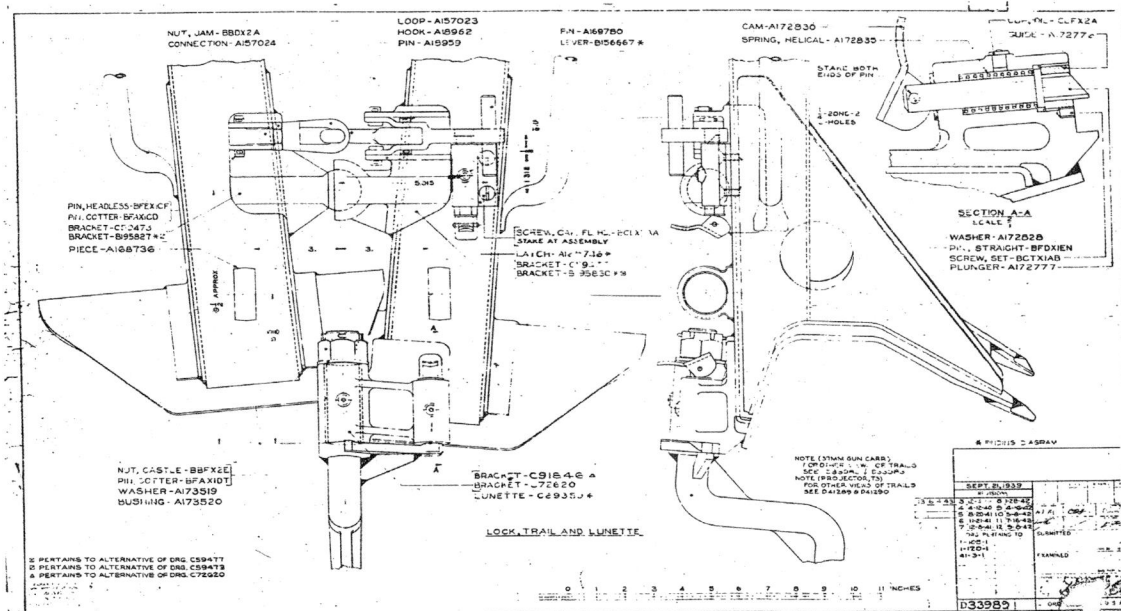

Trail arms lock system and lunette attachment.

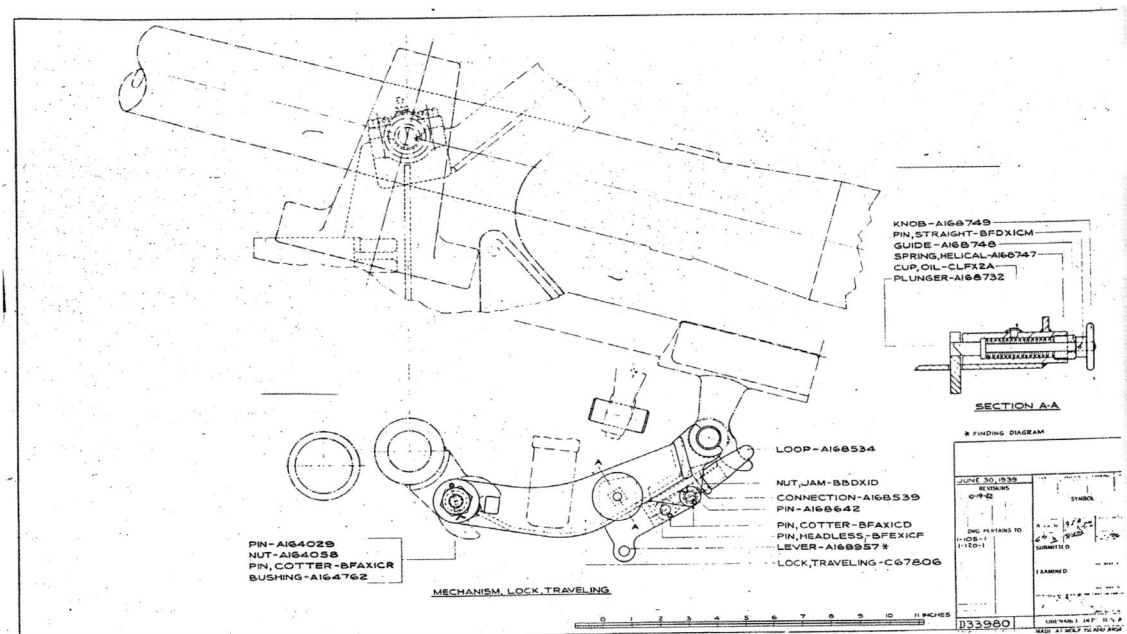

Travel lock hook and engagement lever used when gun is towed.

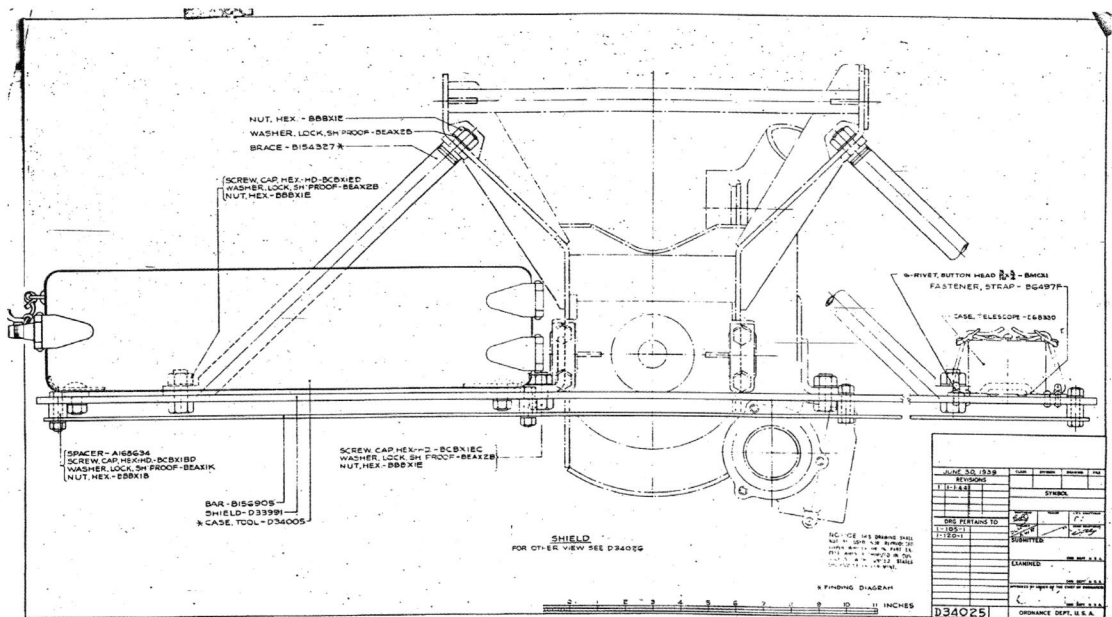

Top view of main armored plate and gun sight box.

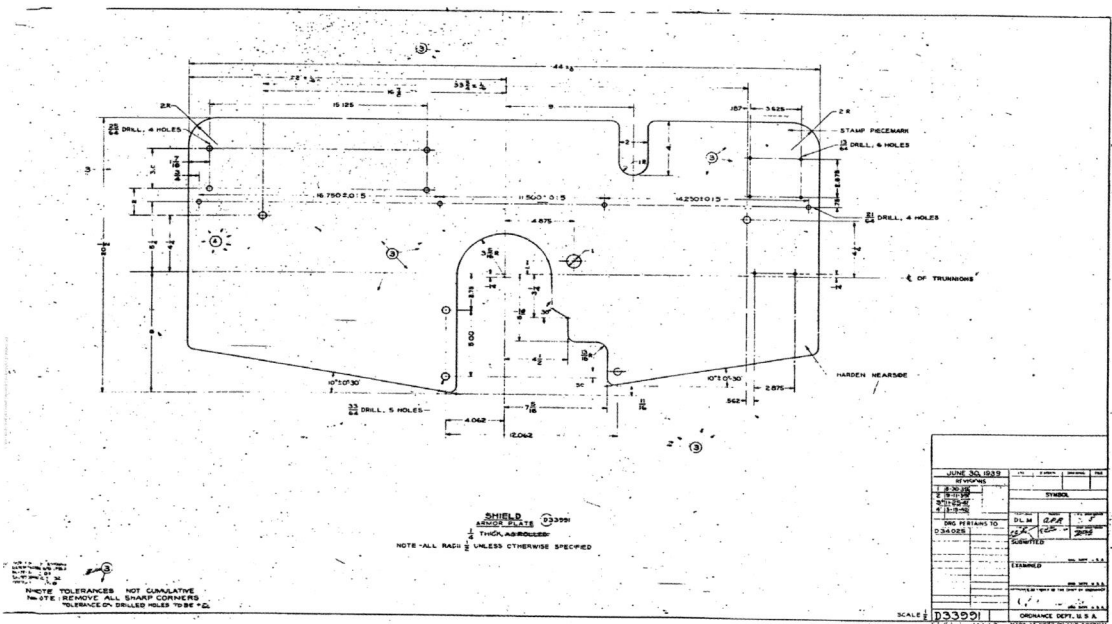

Front view of main armored shield.

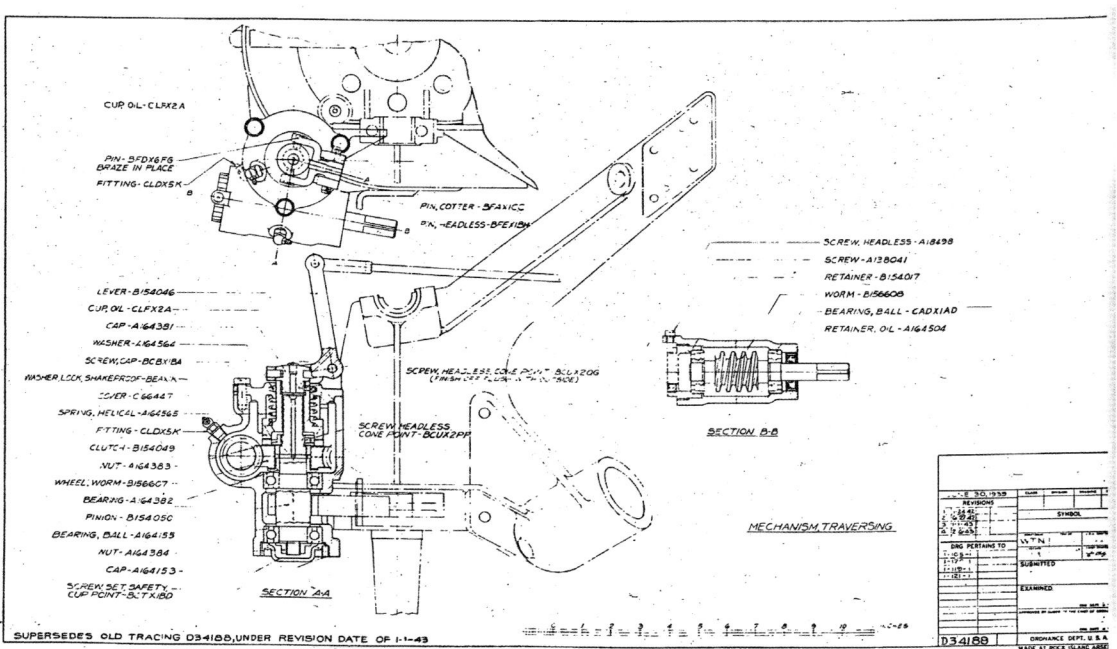

Gun traversing mechanism.

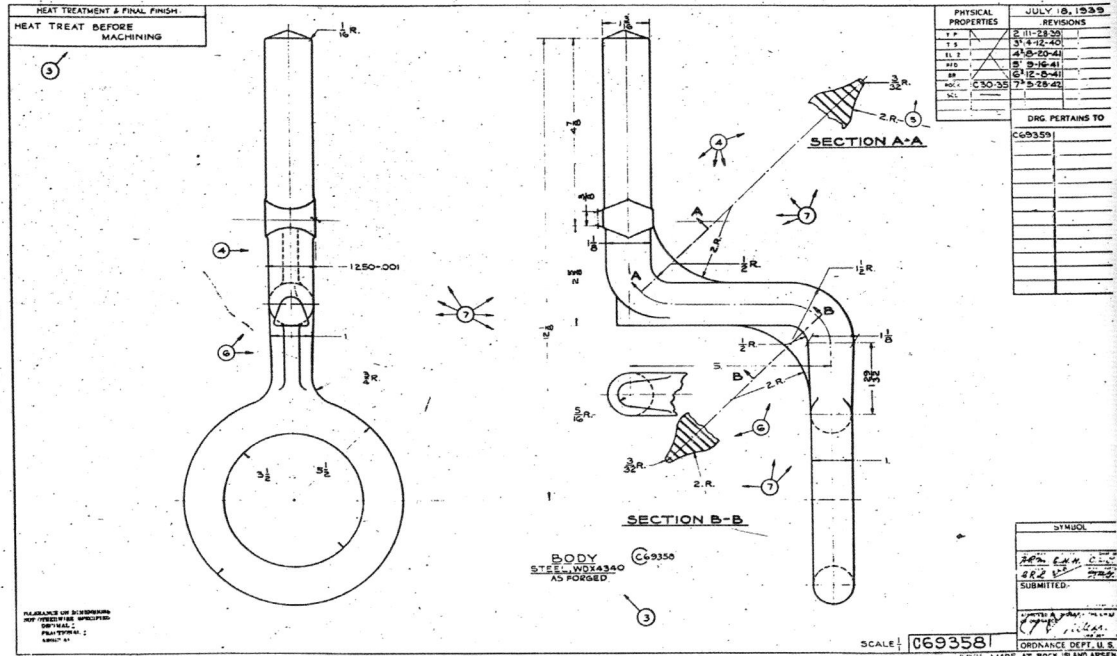

Detailed drawing of the lunette.

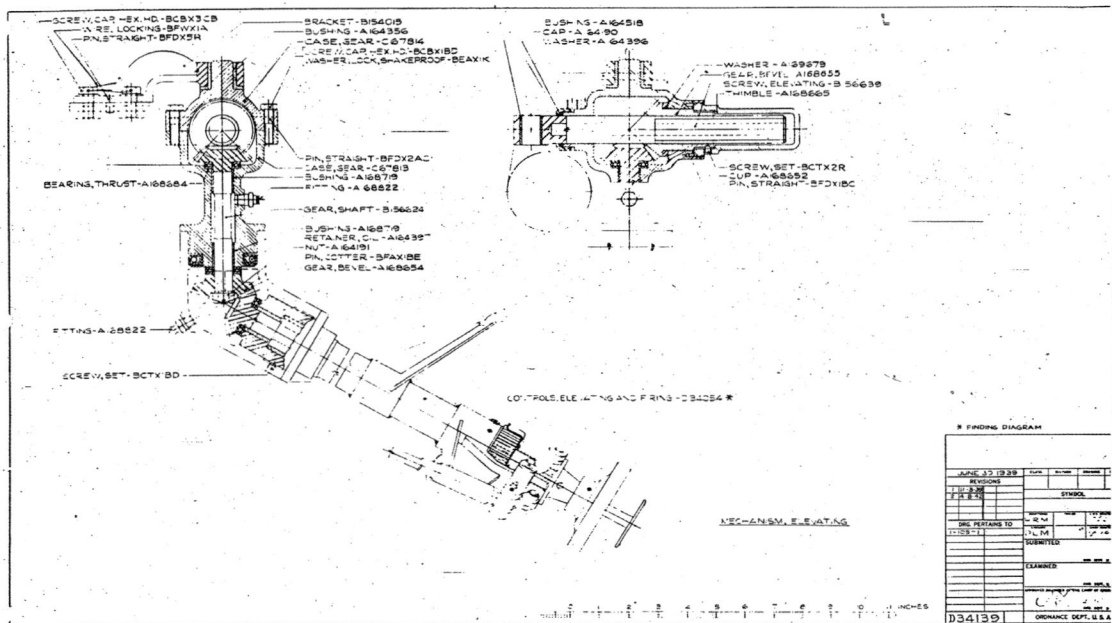

Gun aiming mechanism.

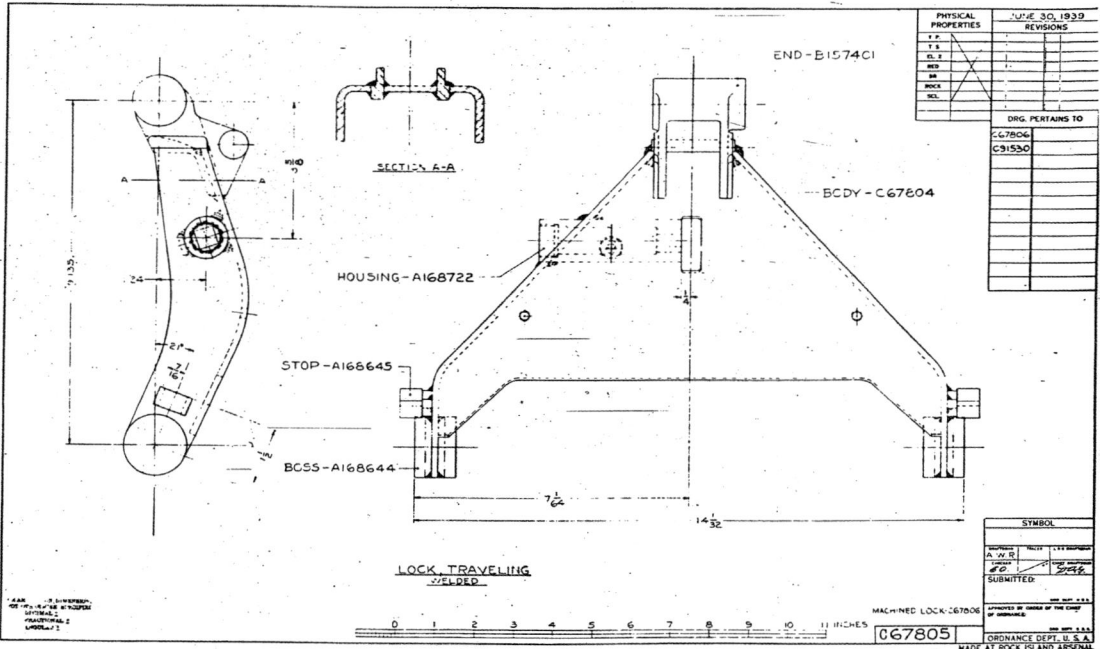

Travel lock plate.

| Appendix II

37-mm Ballistic Data

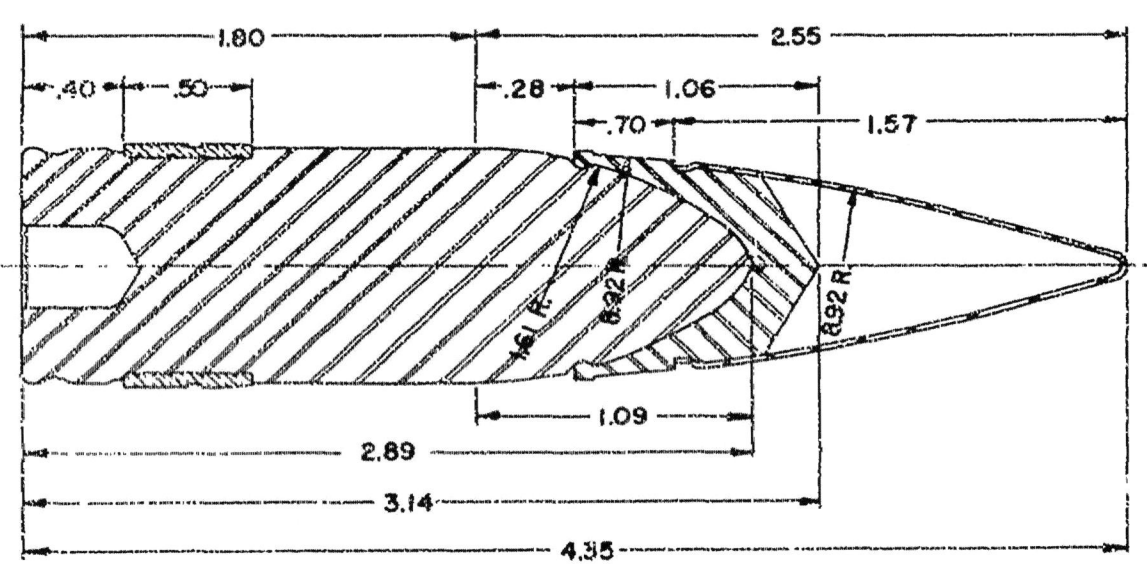

ALL DIMENSIONS IN CALIBERS
I CAL = 1.457"

SHOT, APC, 37-MM, M51

(Handbook of Ballistic and Engineering Data for Ammunition, Volume 1, Ballistic Research Laboratories, Aberdeen Proving Ground, MD, July 1950)

FOR SHOT, APC, 37-MM, M51

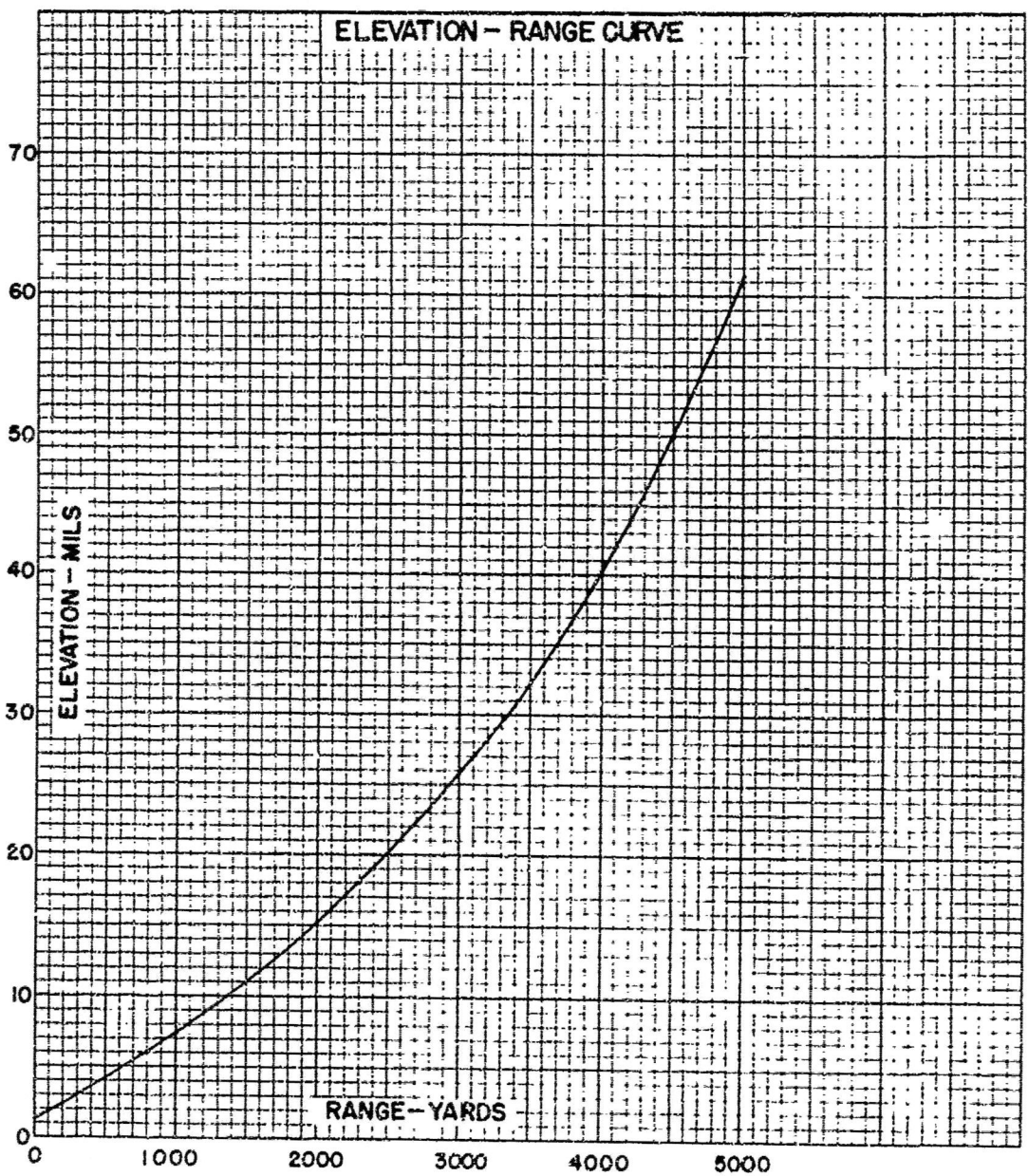

(Handbook of Ballistic and Engineering Data for Ammunition, Volume 1, Ballistic Research Laboratories, Aberdeen Proving Ground, MD, July 1950)

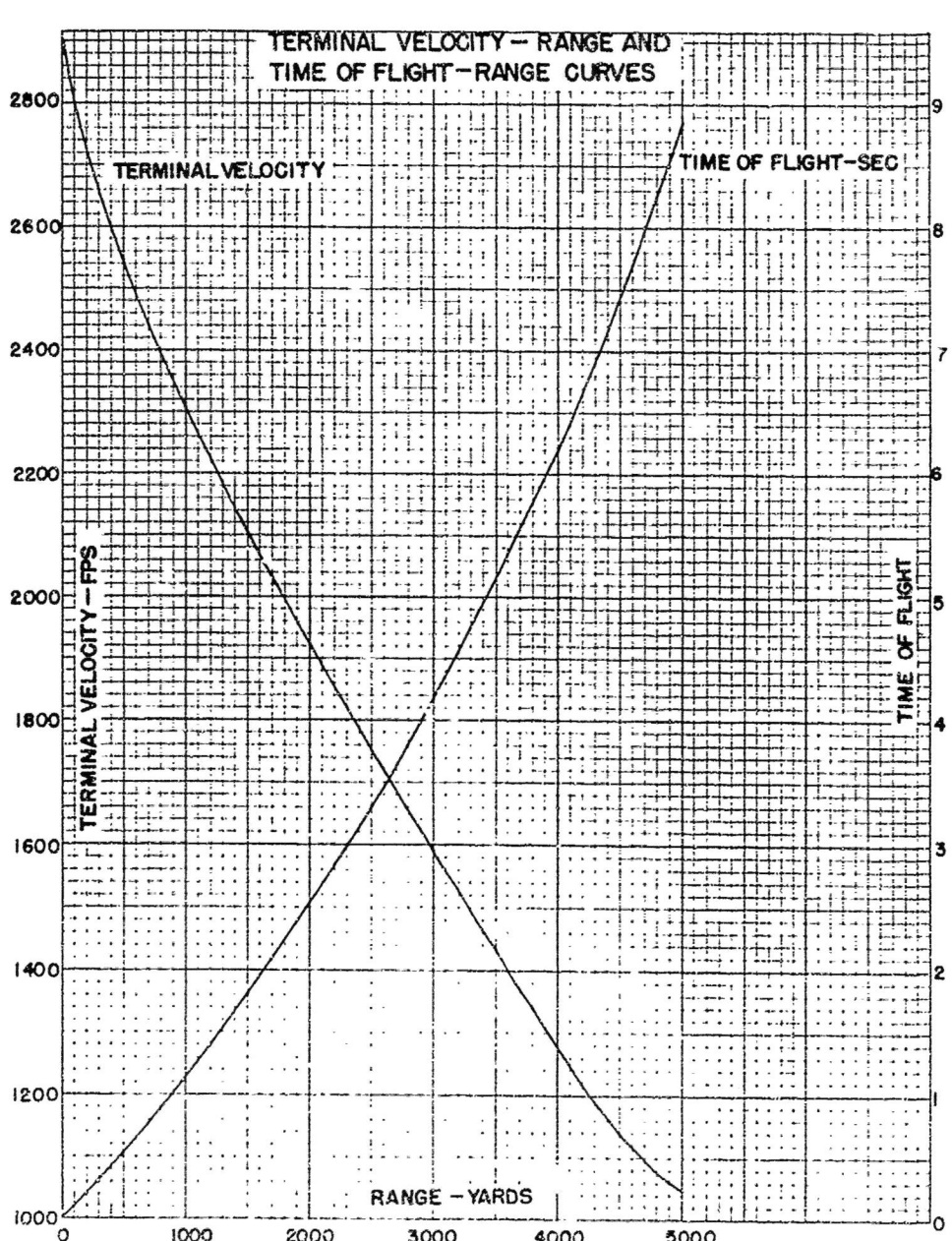

TERMINAL VELOCITY — RANGE AND TIME OF FLIGHT — RANGE CURVES

(Handbook of Ballistic and Engineering Data for Ammunition, Volume 1, Ballistic Research Laboratories, Aberdeen Proving Ground, MD, July 1950)

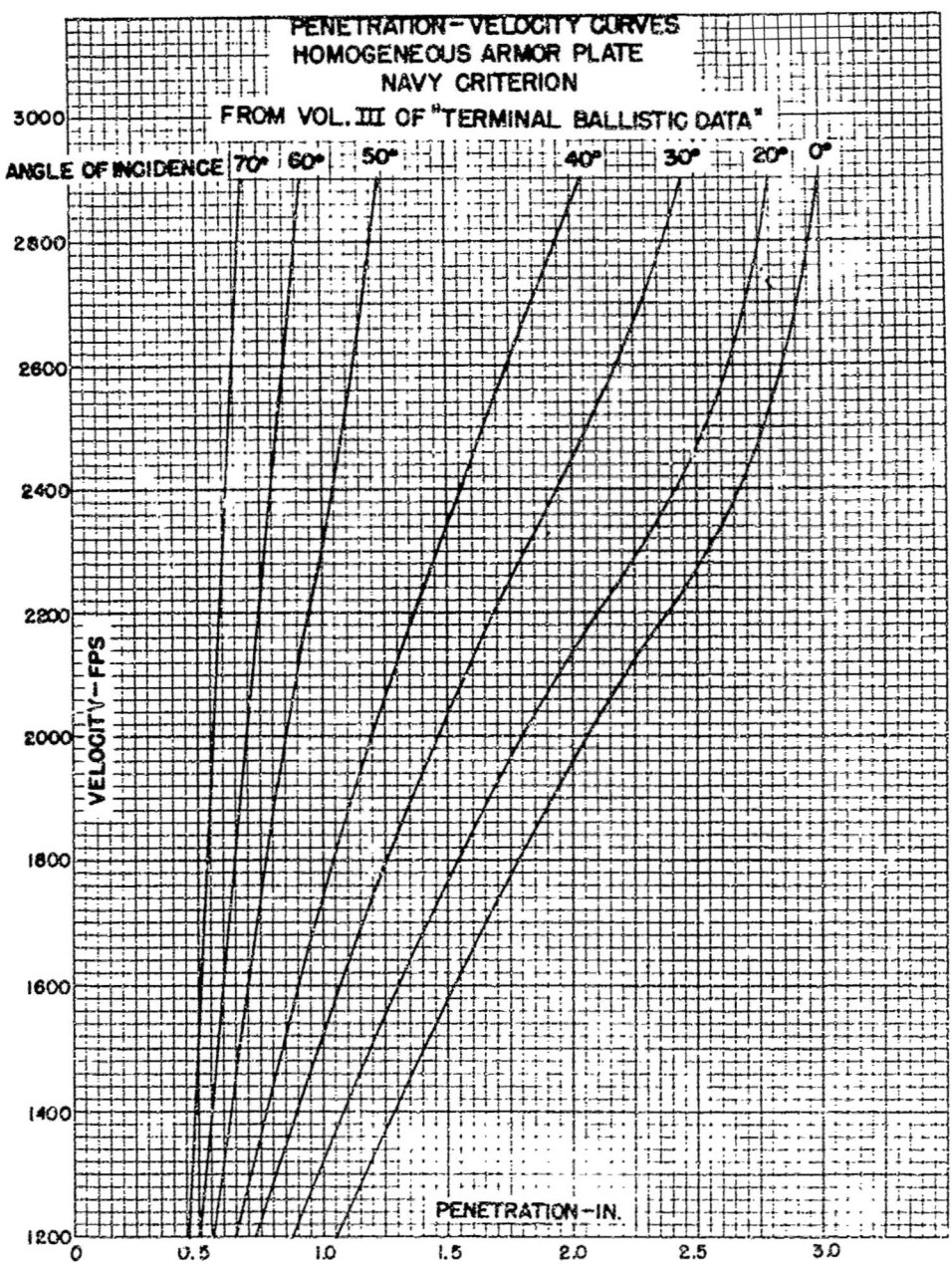

FOR SHOT, APC, 37-MM, M51

PENETRATION–VELOCITY CURVES
HOMOGENEOUS ARMOR PLATE
NAVY CRITERION
FROM VOL. III OF "TERMINAL BALLISTIC DATA"

(Handbook of Ballistic and Engineering Data for Ammunition, Volume 1, Ballistic Research Laboratories, Aberdeen Proving Ground, MD, July 1950)

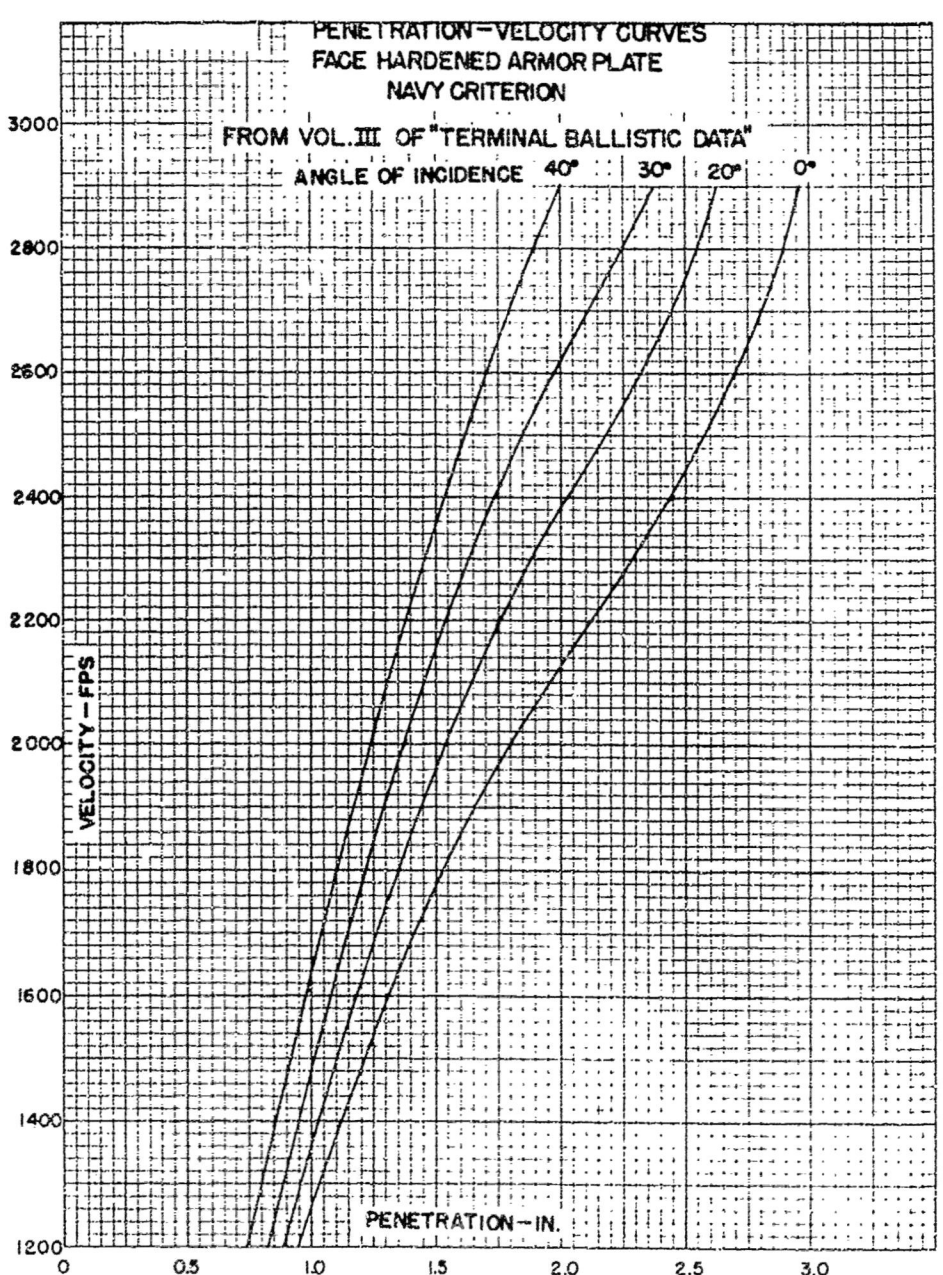

(Handbook of Ballistic and Engineering Data for Ammunition, Volume 1, Ballistic Research Laboratories, Aberdeen Proving Ground, MD, July 1950)

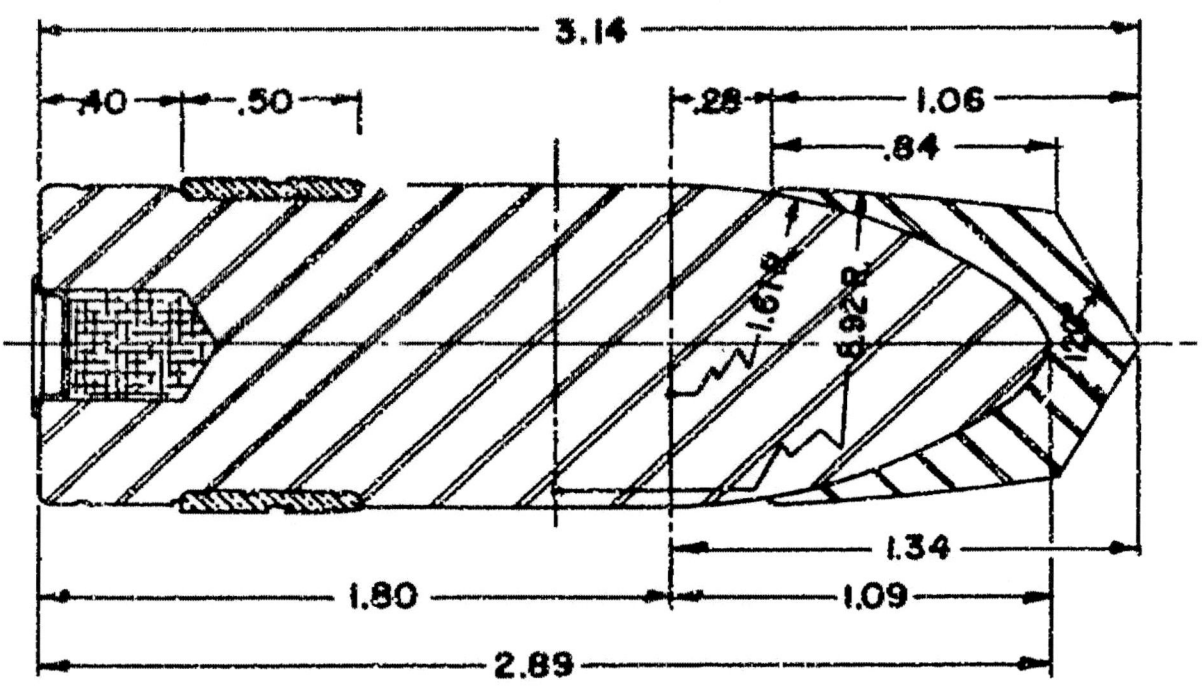

ALL DIMENSIONS IN CALIBERS
I CAL = 1.457"

SHOT, APC, 37-MM, M59

(Handbook of Ballistic and Engineering Data for Ammunition, Volume 1, Ballistic Research
Laboratories, Aberdeen Proving Ground, MD, July 1950)

FOR SHOT, APC, 37-MM, M59

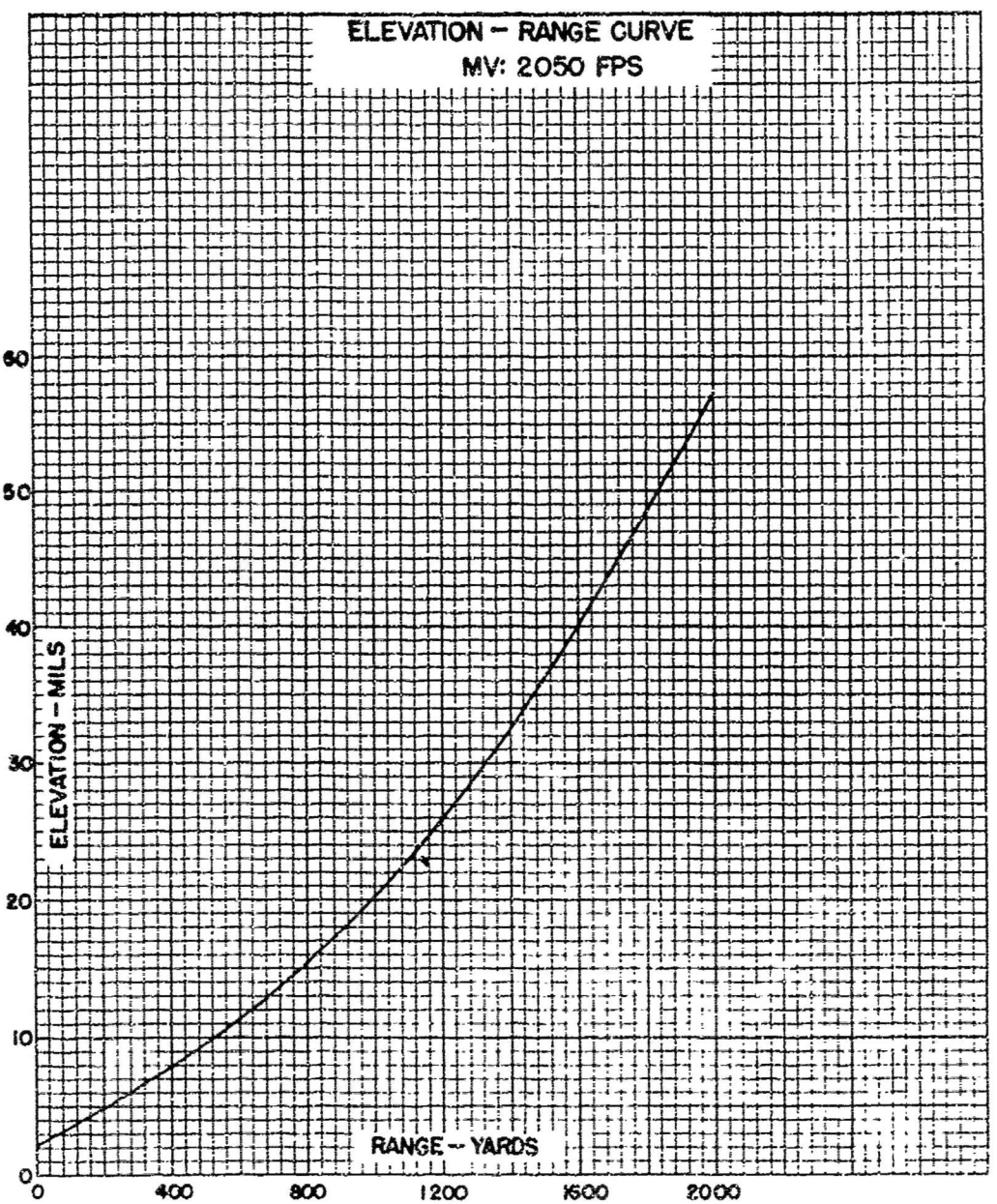

(Handbook of Ballistic and Engineering Data for Ammunition, Volume 1, Ballistic Research Laboratories, Aberdeen Proving Ground, MD, July 1950)

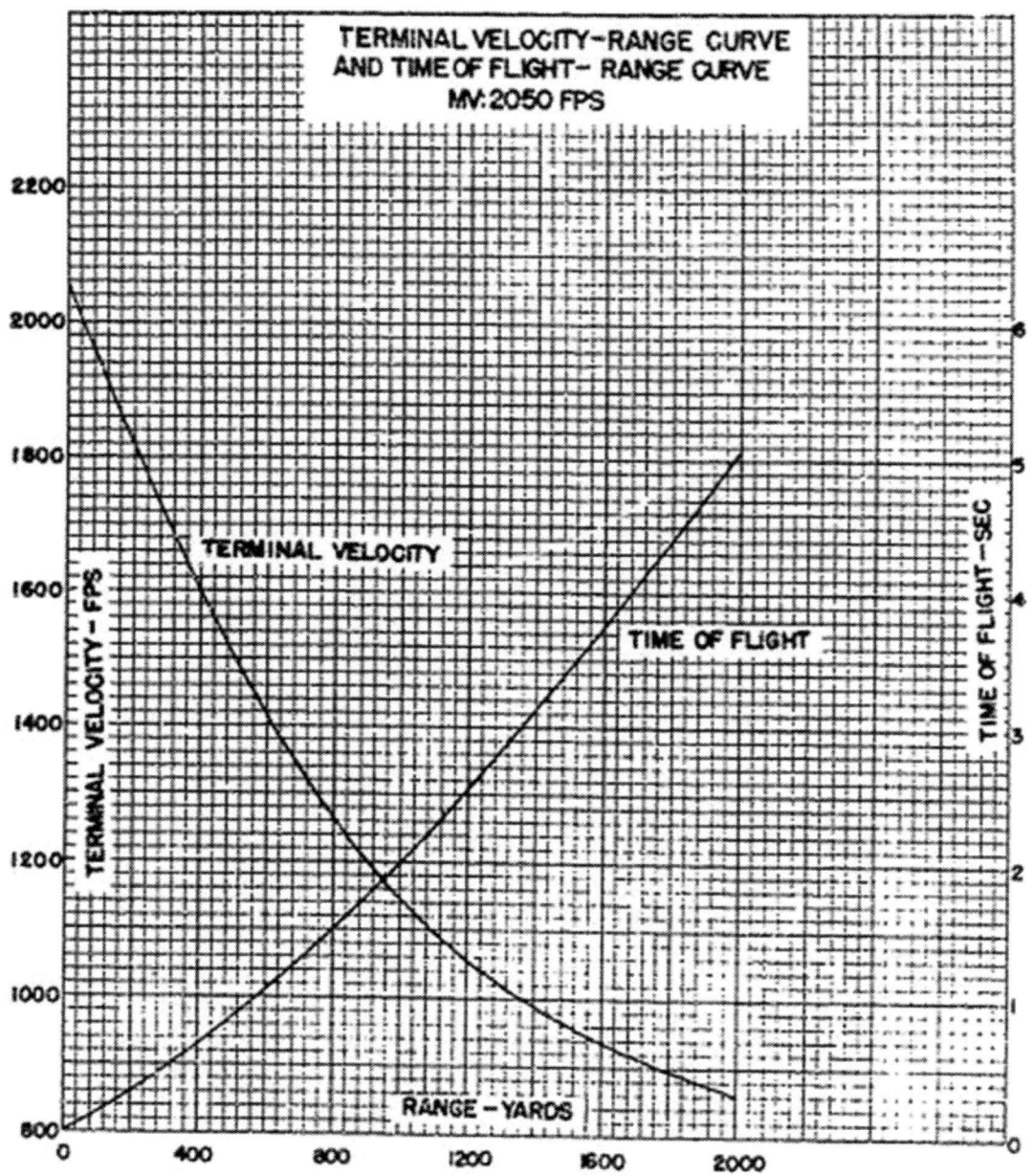

(Handbook of Ballistic and Engineering Data for Ammunition, Volume 1, Ballistic Research Laboratories, Aberdeen Proving Ground, MD, July 1950)

M59

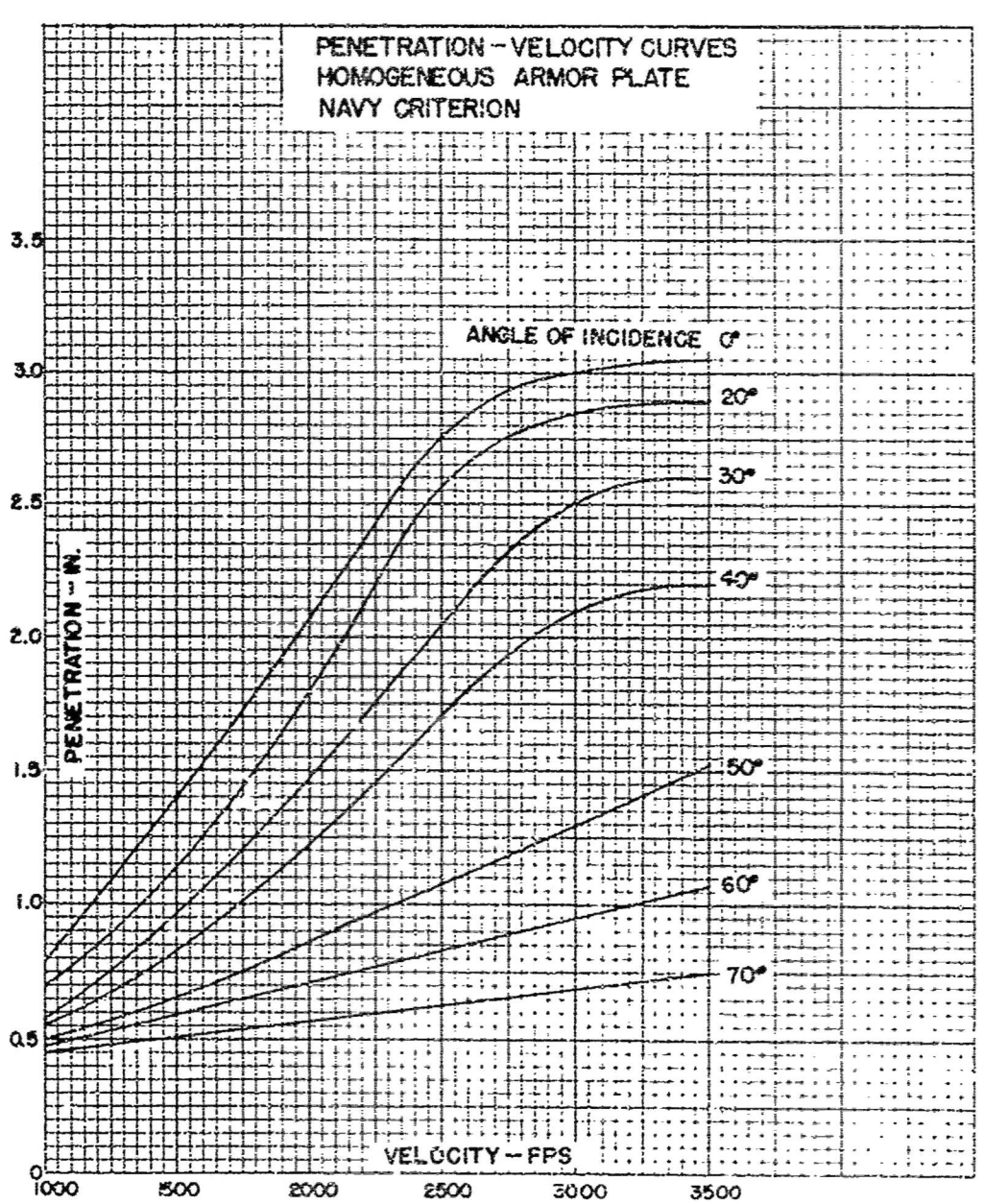

(Handbook of Ballistic and Engineering Data for Ammunition, Volume 1, Ballistic Research Laboratories, Aberdeen Proving Ground, MD, July 1950)

FOR SHOT, APC, 37-MM, M59

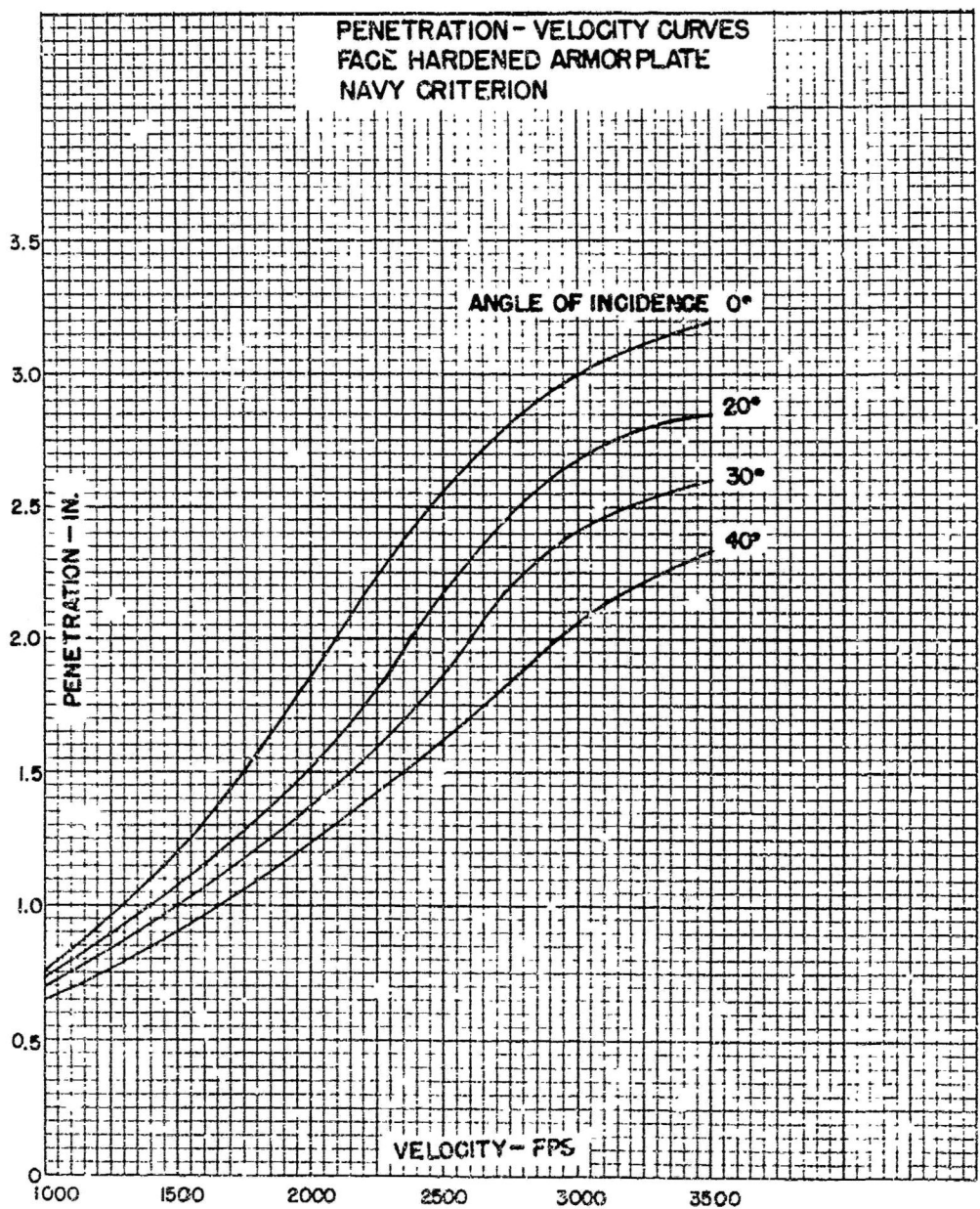

(Handbook of Ballistic and Engineering Data for Ammunition, Volume 1, Ballistic Research Laboratories, Aberdeen Proving Ground, MD, July 1950)

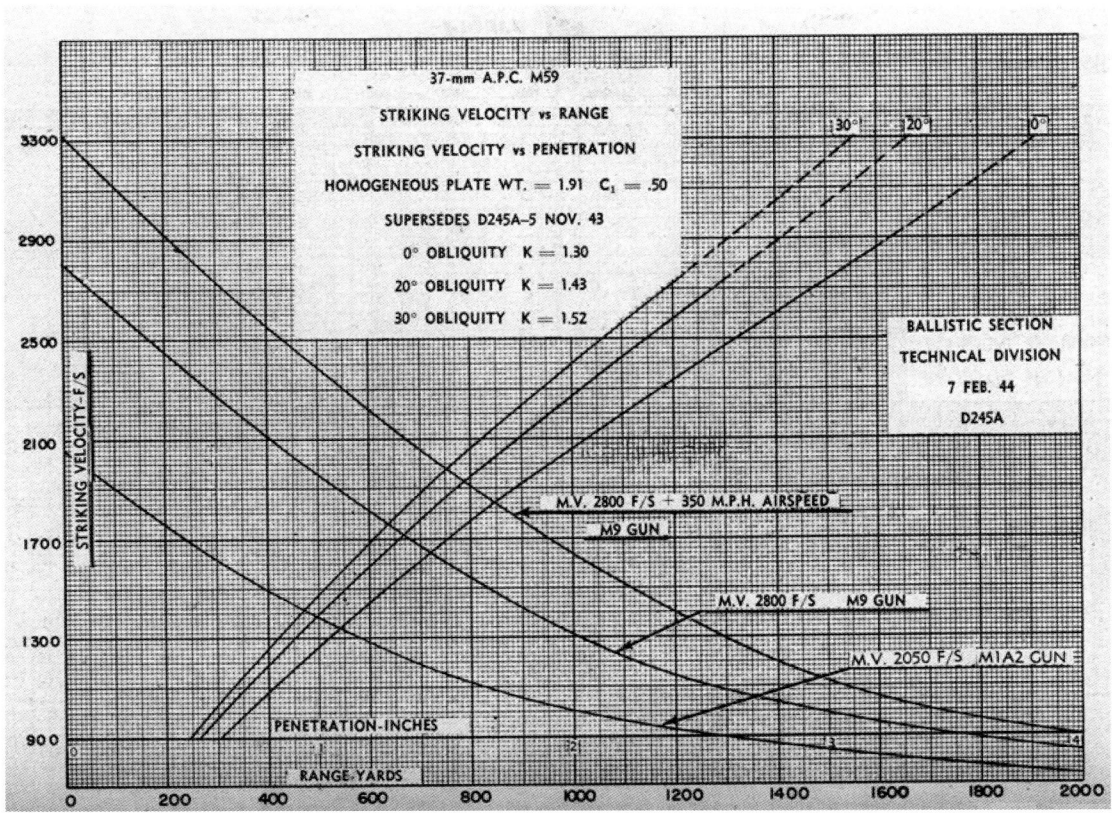

TM9-1907 Ballistic Data Performance of Ammunition, War Department, 23 September 1944

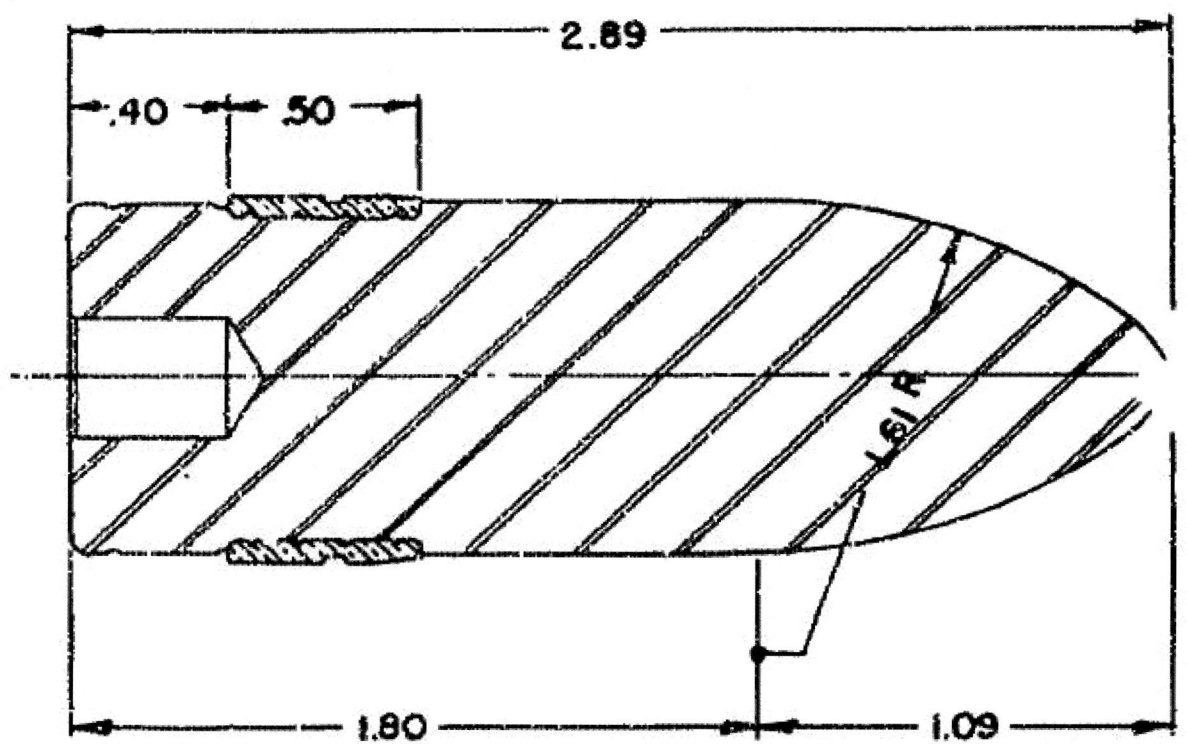

ALL DIMENSIONS IN CALIBERS
1 CAL = 1.457"

SHOT, AP, 37-MM, M80

(Handbook of Ballistic and Engineering Data for Ammunition, Volume 1, Ballistic Research Laboratories, Aberdeen Proving Ground, MD, July 1950)

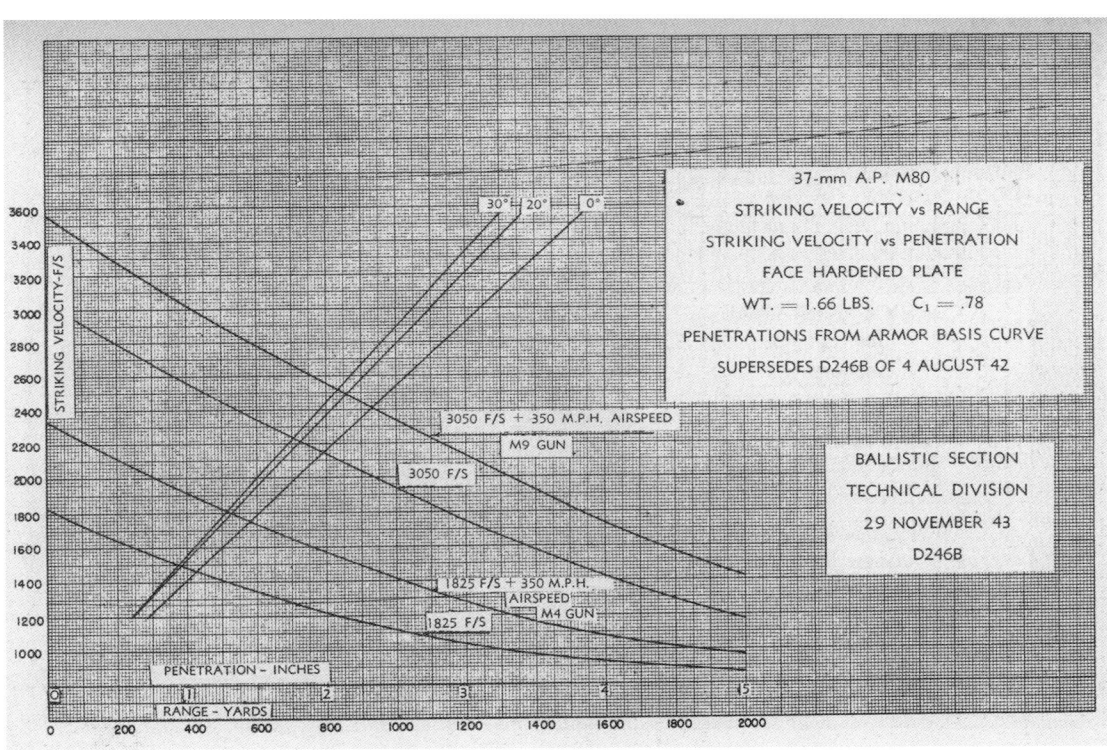

TM9-1907 Ballistic Data Performance of Ammunition, War Department, 23 September 1944

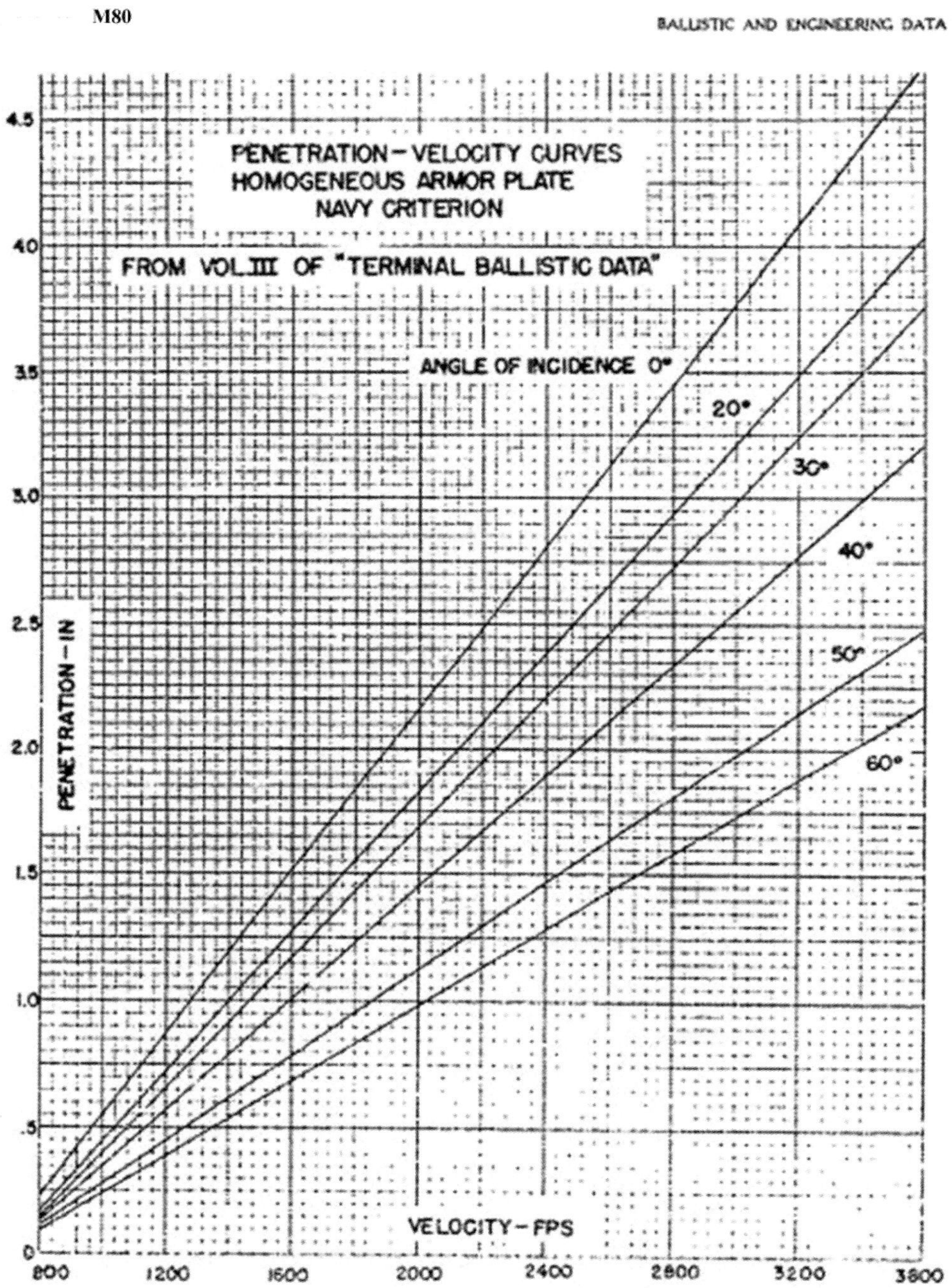

FOR SHOT, AP, 37-MM, M60

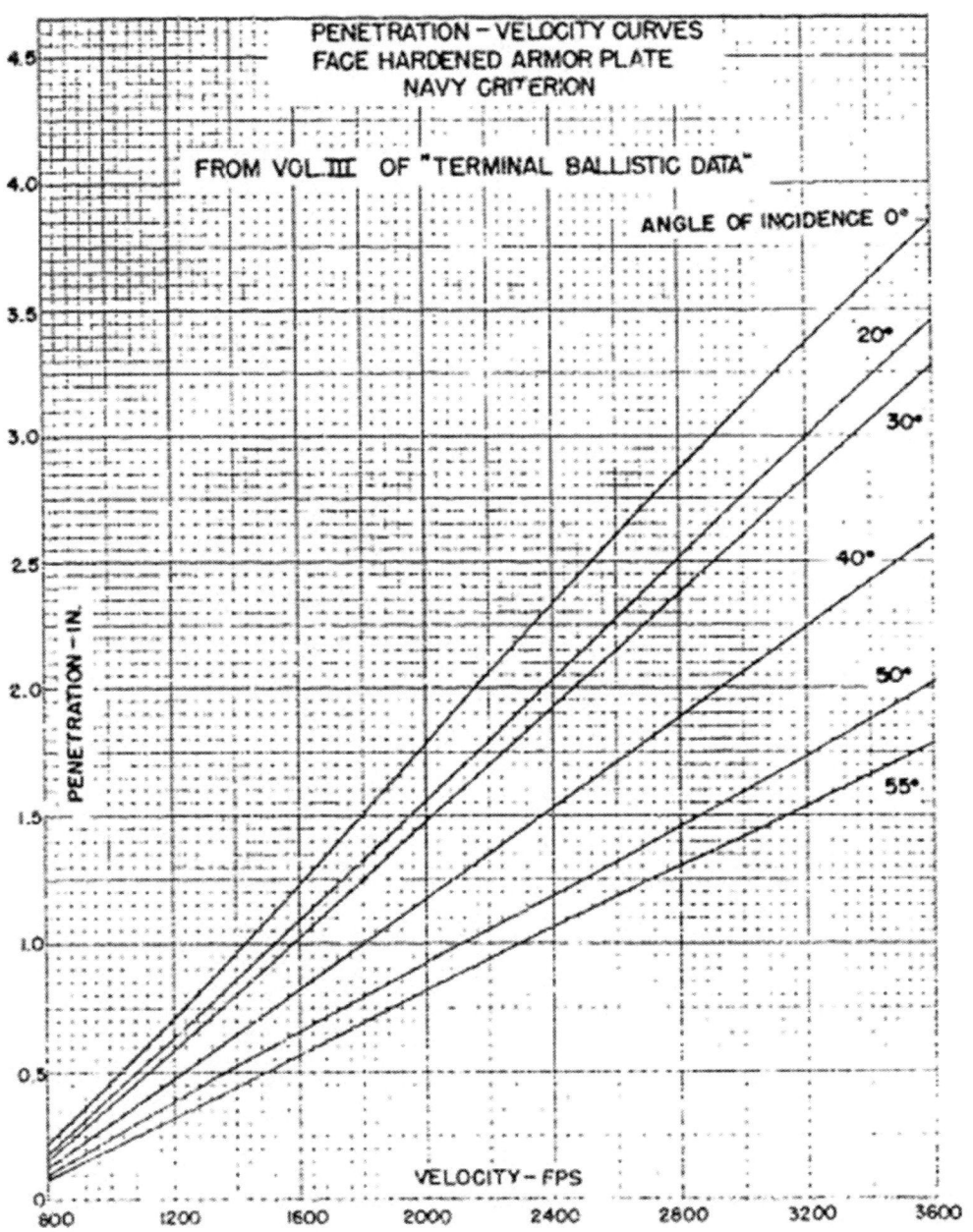

(Handbook of Ballistic and Engineering Data for Ammunition, Volume 1, Ballistic Research Laboratories, Aberdeen Proving Ground, MD, July 1950)

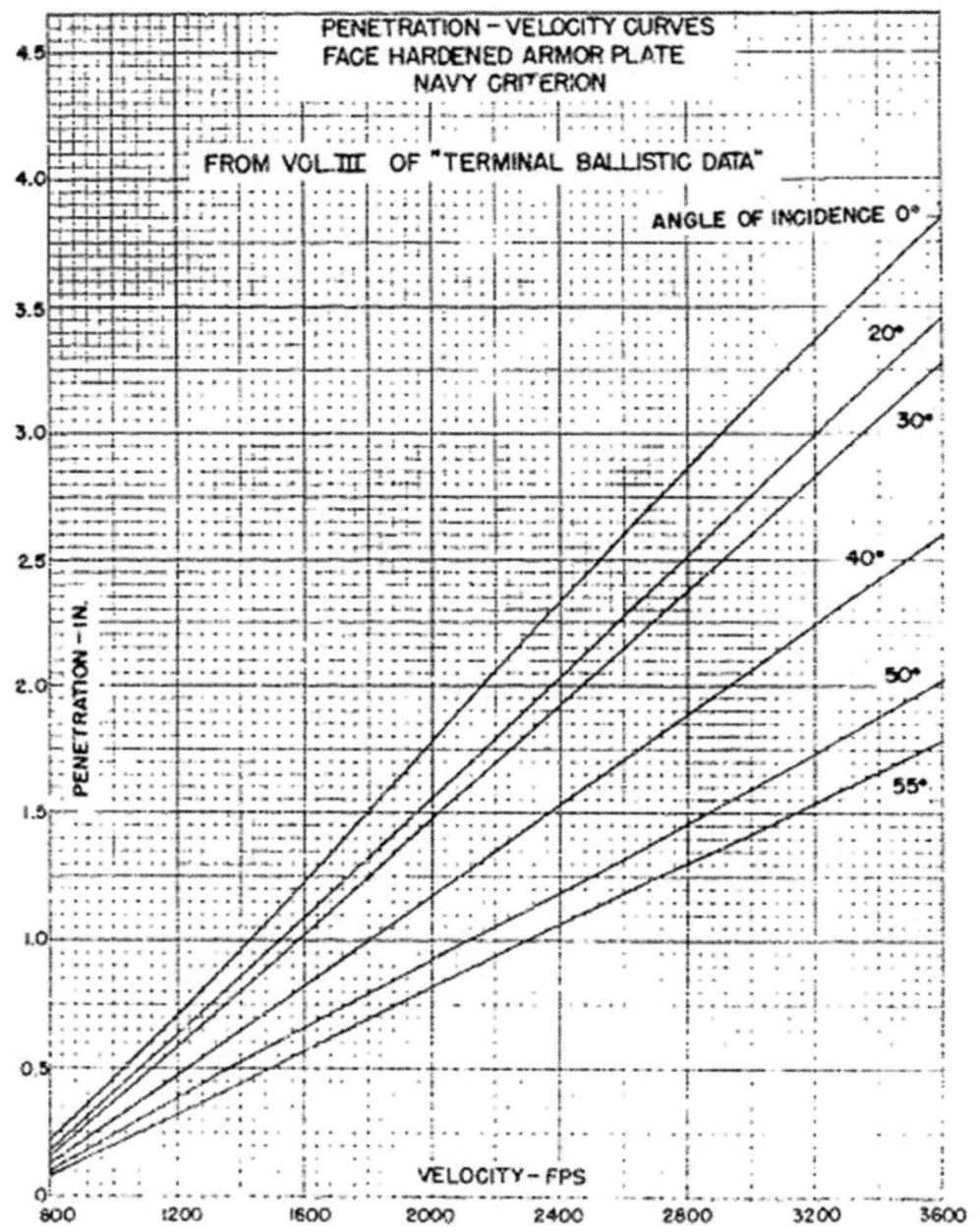

(Handbook of Ballistic and Engineering Data for Ammunition, Volume 1, Ballistic Research Laboratories, Aberdeen Proving Ground, MD, July 1950)

| Further Reading

FM 23-70 Basic Field Manual, 37-mm Gun, Antitank M3, 1940.

FM 23-75 Basic Field Manual, 37-mm Gun 1916.

FM 23-80 Basic Field Manual, 37-mm Gun, Tank, M5, June 25, 1941.

FM 23-81 Basic Field Manual, 37-mm Gun, Tank, M6, April 3, 1942.

Handbook of Ballistic and Engineering Data for Ammunition, Vol. 1, Ballistic Research Laboratories, Aberdeen Proving Ground, MD, July 1950.

Roberts, C., *Armored Strike Force: The Photo History of the 70th Tank Battalion in World War II*, Stackpole, New York, 2016.

Roberts, C., *The Boat That Won the War*, Pen & Sword, South Yorkshire, England 2017.

Roberts, C., *U.S. Airborne Tanks 1939–1945*, Frontline Books, Yorkshire, England, 2021.

TO 11-1-70 37-mm Automatic Aircraft Gun M9, August 1, 1944.

TM 9-1245 Technical Manual, 37-MM Antitank Guns M3 and M3A1 on Carriages M4 and M4A1, War Department, February 1943.

TM 9-1904 Technical Manual, Ammunition Inspection Guide, War Department, 1944.

TM 9-1907 Technical Manual, Ballistic Data, Performance of Ammunition, War Department, September 23, 1944.

TM 9-235 Technical Manual, 37-mm AA Gun Material, War Department, January 24, 1944.

TM 9-240 Technical Manual, 37-mm Automatic Guns AN-M4 and M10, Aircraft, War Department, May 7, 1945.

TM 9-246 Technical Manual, 37-mm Gun T32 and Tripod Mount T9. War Department, August 1944.

TM 9-726 Technical Manual, Light Tank M3, Technical Manual, War Department, July 15, 1942.

TM 9-732 Technical Manual, Light tanks M5 and M5A1, War Department, November 27, 1943.

TM 9-741, Technical Manual, Staghound Medium Armored Car T17E1, War Department, December 15, 1942.

TM 9-743, Technical Manual, Light Armored Car M8 and Armored Utility Car M20, War Department, February 21, 1944.

TM 9-750, Technical Manual, Ordnance Maintenance Lee Medium Tanks, M3, M3A1 and M3A2, War Department, May 9, 1942.

TM 9-750A, Technical Manual, 37-MM Gun Motor Carriage M6, War Department, November 25, 1942.

TM 9-775, Technical Manual, Landing Vehicles Tracked, LVT MKI and MKII, War Department, February 5, 1944.

TM-E 30-451, Handbook on German Military Forces, September 1, 1943.

Index

Other Titles in the Casemate Illustrated Special Series

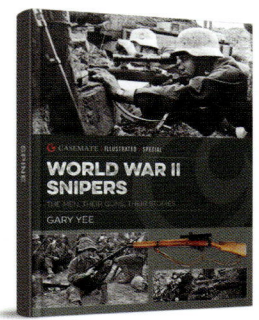

WORLD WAR II SNIPERS

by Gary Yee

A comprehensive illustrated history of snipers and their battlefield role in World War II, using neglected or forgotten sources to illuminate their victories and defeats.

MAY 2022 | 9781636240985

U.S. ARMY SIGNAL CORPS VEHICLES 1941–45

by Didier Andres

A comprehensive and fully illustrated account of all the vehicles needed to move, use, and maintain communications equipment vital to the success of the U.S. Army during World War II.

OCTOBER 2021 | 9781636240640

U.S. ARMY AMBULANCES AND MEDICAL VEHICLES IN WORLD WAR II

by Didier Andres

A fully illustrated book covering all types of ambulances and medical vehicles used by the U.S. Army during World War II.

JULY 2020 | 9781612008653

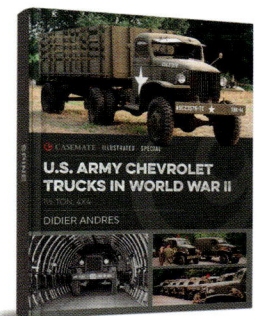

U.S. ARMY CHEVROLET TRUCKS IN WORLD WAR II

by Didier Andres

A fully illustrated and detailed account of the 1 ½-ton Chevy truck and its use by the U.S. Army during World War II.

MAY 2020 | 9781612008639